MongoDB 参考手册(影印版)

Amol Nayak 著

南京　东南大学出版社

图书在版编目(CIP)数据

MongoDB 参考手册:英文/(印)纳亚克(Nayak,
A.)著. —影印本. —南京:东南大学出版社,2016.1
 书名原文:MongoDB Cookbook
 ISBN 978-7-5641-6092-0

Ⅰ.①M… Ⅱ.①纳… Ⅲ.①关系数据库系统-
手册-英文 Ⅳ.①TP311.138-62

中国版本图书馆 CIP 数据核字(2015)第 256608 号

© 2014 by PACKT Publishing Ltd

Reprint of the English Edition, jointly published by PACKT Publishing Ltd and Southeast University Press, 2016. Authorized reprint of the original English edition, 2015 PACKT Publishing Ltd, the owner of all rights to publish and sell the same.

All rights reserved including the rights of reproduction in whole or in part in any form.

英文原版由 PACKT Publishing Ltd 出版 2014。

英文影印版由东南大学出版社出版 2016。此影印版的出版和销售得到出版权和销售权的所有者——PACKT Publishing Ltd 的许可。

版权所有,未得书面许可,本书的任何部分和全部不得以任何形式重制。

MongoDB 参考手册(影印版)

出版发行:	东南大学出版社
地 址:	南京四牌楼 2 号 邮编:210096
出 版 人:	江建中
网 址:	http://www.seupress.com
电子邮件:	press@seupress.com
印 刷:	常州市武进第三印刷有限公司
开 本:	787 毫米×980 毫米 16 开本
印 张:	24.25
字 数:	475 千字
版 次:	2016 年 1 月第 1 版
印 次:	2016 年 1 月第 1 次印刷
书 号:	ISBN 978-7-5641-6092-0
定 价:	74.00 元

本社图书若有印装质量问题,请直接与营销部联系。电话(传真):025-83791830

Credits

Author
Amol Nayak

Reviewers
Jan Borgelin
Doug Duncan
Laurence Putra
Liran Tal
Khaled Tannir

Acquisition Editor
Neha Nagwekar

Content Development Editor
Priyanka Shah

Technical Editors
Veronica Fernandes
Ankita Thakur

Copy Editors
Karuna Narayanan
Shambhavi Pai

Project Coordinators
Mary Alex
Neha Thakur

Proofreaders
Stephen Copestake
Paul Hindle
Kelly Hutchinson
Clyde Jenkins

Indexers
Mariammal Chettiyar
Monica Ajmera Mehta
Rekha Nair

Graphics
Sheetal Aute
Abhinash Sahu

Production Coordinators
Kyle Albuquerque
Conidon Miranda
Nitesh Thakur

Cover Work
Kyle Albuquerque

About the Author

Amol Nayak is a certified MongoDB developer and has been working as a developer for over 8 years. He is currently employed with a leading financial data provider, working on cutting-edge technologies. He has used MongoDB as a database for various systems at his current and previous workplaces to support enormous data volumes. He is an open source enthusiast and supports it by contributing to open source frameworks and promoting them. He has made contributions to the Spring Integration project, and his contributions are adapters for JPA, XQuery, and MongoDB and push notifications for mobile devices and Amazon Web Services (AWS). He also has some contributions to the Spring Data MongoDB project. Apart from technology, he is passionate about motor sports and is a race official at Buddh International Circuit, India, for various motor-sports events. Earlier, he was the author of *Instant MongoDB, Packt Publishing*.

> I would like to thank everyone at Packt Publishing who have been involved with this book. It started when Luke Presland from Packt Publishing approached me to author a book on MongoDB. I was skeptical to take up the opportunity due to other commitments and tight deadlines, but if it wasn't for my mom, friends, and office colleagues, who convinced me to take up the opportunity, I would not have written this book. The chapters and content to be covered was a lot, and I was having a tough time keeping up with the timelines. A special thanks to Priyanka Shah, Rebecca Pedley, Mary Alex, and Joel Goveya, with whom I interacted the most; they were very flexible to my changes in delivery timelines. A big thanks to Doug Duncan and other reviewers of the book for reviewing the book closely and helping improve the quality of the content drastically. Finally, I would like to thank the other staff at Packt Publishing who were involved in the book's publishing process but haven't interacted with me.

About the Reviewers

Jan Borgelin is a technical geek with over 15 years of professional software-development experience. He is currently the CTO of BA Group Ltd., a consultancy based in Finland. BA Group was one of the early adopters of MongoDB and the first official MongoDB partner in Scandinavia.

Doug Duncan has been working with RDBMSes for the past 15 years and has started shifting gears towards the newer data stores since the past 3 years. He has focused mainly on MongoDB since he came across the 0.8 release. In addition to his day job as a MongoDB database administrator, he works as an online teaching assistant for the MongoDB education team for several of their online courses (https://university.mongodb.com/), where he helps students understand how MongoDB works. When not working, he likes to read about new technologies and try to figure out how they can integrate and work in conjunction with the more established systems already in place.

Laurence Putra is a software engineer working in Singapore and runs the Singapore MongoDB User Group. In his free time, he hacks away on random stuff and picks up new technologies. His key interests lie in security and distributed systems. For more information, view his profile at geeksphere.net.

Liran Tal is a certified MongoDB developer and top contributor to the open source MEAN.IO and MEAN.JS full-stack JavaScript frameworks. Being an avid supporter of and contributor to the open source movement, in 2007, he redefined network RADIUS management by establishing daloRADIUS, a world-recognized and industry-leading open source project.

Liran is currently working at HP Software as an R&D team leader on a combined technology stack featuring a Drupal-based collaboration platform, Java, Node.js, and MongoDB.

At HP Live Network, Liran plays a key role in system-architecture design, shaping the technology strategy from planning and development to deployment and maintenance in HP's IaaS cloud. Acting as the technological focal point, he loves mentoring team mates, driving for better code methodology, and seeking out innovative solutions to support business strategies.

He has a *cum laude* (honors) in his Bachelor's degree in Business and Information Systems Analysis studies and enjoys spending his time with his beloved wife, Tal, and his new born son, Ori. Among other things, his hobbies include playing the guitar, hacking all things on Linux, and continuously experimenting with and contributing to open source projects.

Khaled Tannir is a visionary solution architect with more than 20 years of technical experience, focusing on Big Data technologies and data mining since 2010.

He is widely recognized as an expert in these fields and has a Master of Research degree in Big Data and Cloud Computing and a Master's degree in System Information Architectures with initially a Bachelor of Technology degree in Electronics.

Khaled is a Microsoft Certified Solutions Developer (MCSD) and an avid technologist. He worked for many companies in France (and recently in Canada), leading the development and implementation of software solutions and giving technical presentations.

He is the author of *RavenDB 2.x Beginner's Guide* and *Optimizing Hadoop for MapReduce* and is the technical reviewer for *Pentaho Analytics for MongoDB* and *MongoDB High Availability*, all available at Packt Publishing.

He enjoys taking landscape and night photos; traveling; playing video games; creating funny electronic gadgets with Arduino, Raspberry PI, and .NET Gadgeteer; and of course, spending time with his wife and family.

You can reach him at contact@khaledtannir.net.

www.PacktPub.com

Support files, eBooks, discount offers, and more

For support files and downloads related to your book, please visit www.PacktPub.com.

Did you know that Packt offers eBook versions of every book published, with PDF and ePub files available? You can upgrade to the eBook version at www.PacktPub.com and as a print book customer, you are entitled to a discount on the eBook copy. Get in touch with us at service@packtpub.com for more details.

At www.PacktPub.com, you can also read a collection of free technical articles, sign up for a range of free newsletters and receive exclusive discounts and offers on Packt books and eBooks.

https://www2.packtpub.com/books/subscription/packtlib

Do you need instant solutions to your IT questions? PacktLib is Packt's online digital book library. Here, you can search, access, and read Packt's entire library of books.

Why Subscribe?

- Fully searchable across every book published by Packt
- Copy and paste, print, and bookmark content
- On demand and accessible via a web browser

Free Access for Packt account holders

If you have an account with Packt at www.PacktPub.com, you can use this to access PacktLib today and view 9 entirely free books. Simply use your login credentials for immediate access.

Table of Contents

Preface — 1

Chapter 1: Installing and Starting the MongoDB Server — 7
- Introduction — 7
- Single node installation of MongoDB — 8
- Starting a single node instance using command-line options — 9
- Single node installation of MongoDB with options from the config file — 12
- Connecting to a single node from the Mongo shell with a preloaded JavaScript — 13
- Connecting to a single node from a Java client — 15
- Starting multiple instances as part of a replica set — 21
- Connecting to the replica set from the shell to query and insert data — 26
- Connecting to the replica set to query and insert data from a Java client — 29
- Starting a simple sharded environment of two shards — 32
- Connecting to a shard from the Mongo shell and performing operations — 37

Chapter 2: Command-line Operations and Indexes — 41
- Creating test data — 41
- Performing simple querying, projections, and pagination from the Mongo shell — 43
- Updating and deleting data from the shell — 46
- Creating an index and viewing plans of queries — 48
- Background and foreground index creation from the shell — 53
- Creating unique indexes on collection and deleting the existing duplicate data automatically — 57
- Creating and understanding sparse indexes — 59
- Expiring documents after a fixed interval using the TTL index — 63
- Expiring documents at a given time using the TTL index — 66

Table of Contents

Chapter 3: Programming Language Drivers — 69
- Introduction — 69
- Installing PyMongo — 70
- Executing query and insert operations using PyMongo — 73
- Executing update and delete operations using PyMongo — 78
- Aggregation in Mongo using PyMongo — 86
- MapReduce in Mongo using PyMongo — 87
- Executing query and insert operations using a Java client — 90
- Executing update and delete operations using a Java client — 93
- Aggregation in Mongo using a Java client — 98
- MapReduce in Mongo using a Java client — 100

Chapter 4: Administration — 103
- Renaming a collection — 104
- Viewing collection stats — 106
- Viewing database stats — 110
- Disabling the preallocation of data files — 114
- Manually padding a document — 114
- Understanding the mongostat and mongotop utilities — 117
- Estimating the working set — 122
- Viewing and killing the currently executing operations — 124
- Using profiler to profile operations — 129
- Setting up users in MongoDB — 133
- Understanding interprocess security in MongoDB — 141
- Modifying collection behavior using the collMod command — 143
- Setting up MongoDB as a Windows Service — 145
- Configuring a replica set — 147
- Stepping down as a primary instance from the replica set — 155
- Exploring the local database of a replica set — 156
- Understanding and analyzing oplogs — 159
- Building tagged replica sets — 163
- Configuring the default shard for nonsharded collections — 168
- Manually splitting and migrating chunks — 170
- Performing domain-driven sharding using tags — 172
- Exploring the config database in a sharded setup — 175

Chapter 5: Advanced Operations — 179
- Introduction — 180
- Atomic find and modify operations — 180
- Implementing atomic counters in MongoDB — 183
- Implementing server-side scripts — 185
- Creating and tailing capped collection cursors in MongoDB — 188

Converting a normal collection to a capped collection	191
Storing binary data in MongoDB	192
Storing large data in MongoDB using GridFS	194
Storing data to GridFS from a Java client	198
Storing data to GridFS from a Python client	202
Implementing triggers in MongoDB using oplog	204
Executing flat plane (2D) geospatial queries in Mongo using geospatial indexes	209
Spherical indexes and GeoJSON-compliant data in MongoDB	212
Implementing a full-text search in MongoDB	218
Integrating MongoDB with Elasticsearch for a full-text search	224

Chapter 6: Monitoring and Backups — 231

Introduction	231
Signing up for MMS and setting up the MMS monitoring agent	232
Managing users and groups on the MMS console	236
Monitoring MongoDB instances on MMS	239
Setting up monitoring alerts on MMS	248
Backing up and restoring data in Mongo using out-of-the box tools	250
Configuring the MMS backup service	253
Managing backups in the MMS backup service	259

Chapter 7: Cloud Deployment on MongoDB — 267

Introduction	267
Setting up and managing the MongoLab account	268
Setting up a sandbox MongoDB instance on MongoLab	270
Performing operations on MongoDB from MongoLab GUI	274
Setting up MongoDB on Amazon EC2 using the MongoDB AMI	278
Setting up MongoDB on Amazon EC2 without using the MongoDB AMI	289

Chapter 8: Integration with Hadoop — 293

Introduction	293
Executing our first sample MapReduce job using the mongo-hadoop connector	294
Writing our first Hadoop MapReduce job	301
Running MapReduce jobs on Hadoop using streaming	304
Running a MapReduce job on Amazon EMR	308

Chapter 9: Open Source and Proprietary Tools — 317

Introduction	317
Developing using spring-data-mongodb	318
Accessing MongoDB using Java Persistence API	329
Accessing MongoDB over REST	333
Installing the GUI-based client, MongoVUE, for MongoDB	338

Table of Contents

Appendix: Concepts for Reference — 351
Write concern and its significance — 351
Read preference for querying — 360
Index — 365

Preface

MongoDB is a document-oriented, leading NoSQL database, which offers linear scalability, thus making it a good contender for high-volume, high-performance systems across all business domains. It has an edge over the majority of NoSQL solutions for its ease of use, high performance, and rich features.

This book provides detailed recipes that describe how to use the different features of MongoDB. The recipes cover topics ranging from setting up MongoDB, knowing its programming-language API, monitoring and administration, to some advanced topics such as cloud deployment, integration with Hadoop, and some open source and proprietary tools for MongoDB. The recipe format presents the information in a concise, actionable form; this lets you refer to the recipe to address and know the details of just the use case in hand, without going through the entire book.

What this book covers

Chapter 1, Installing and Starting the MongoDB Server, is all about starting MongoDB. It will demonstrate how to start the server in the standalone mode, as a replica set, and as a shard, with the provided start-up options from the command line or config file.

Chapter 2, Command-line Operations and Indexes, has simple recipes to perform CRUD operations from the Mongo shell and create various types of indexes from the shell.

Chapter 3, Programming Language Drivers, is about programming language APIs. Though Mongo supports a vast array of languages, we will look at how to use the drivers to connect to the MongoDB server from Java and Python programs only. This chapter also explores the MongoDB wire protocol used for communication between the server and the programming-language clients.

Chapter 4, Administration, contains many recipes around administration or your MongoDB deployment. This chapter covers a lot of frequently used administrative tasks such as viewing the stats of the collections and database, viewing and killing long-running operations and other replica, and sharding-related administration.

Preface

Chapter 5, *Advanced Operations*, is an extension of *Chapter 2*, *Command-line Operations and Indexes*. We will look at some of the slightly advanced features such as implementing server-side scripts, geospatial search, GridFS, full-text search, and how to integrate MongoDB with an external full-text search engine.

Chapter 6, *Monitoring and Backups*, is all about administration and some basic monitoring. However, MongoDB provides a state-of-the-art monitoring and real-time backup service, MongoDB Monitoring Service (MMS). In this chapter, we will look at some recipes around monitoring and backup using MMS.

Chapter 7, *Cloud Deployment on MongoDB*, covers recipes that use MongoDB service providers for cloud deployment, and we will set up our own MongoDB server on the AWS cloud.

Chapter 8, *Integration with Hadoop*, covers recipes to integrate MongoDB with Hadoop to use the Hadoop MapReduce API to run MapReduce jobs on data residing in MongoDB/MongoDB data files and write the results back to them. We will also see how to use AWS EMR to run our MapReduce jobs on the cloud using Amazon's managed Hadoop cluster, EMR with the mongo-hadoop connector.

Chapter 9, *Open Source and Proprietary Tools*, is about using frameworks and products built around MongoDB to improve a developer's productivity or about making some of the day-to-day jobs in using Mongo easy. Unless explicitly mentioned, the products/frameworks we will be looking at in this chapter are open source.

Appendix, *Concepts for Reference*, gives you a bit of additional information on write concern and read preference for reference.

What you need for this book

The version of MongoDB used to try out the recipes is 2.4.6. The recipes hold good for version 2.6.x as well. In case of some special feature specific to version 2.6.x, it would be explicitly mentioned in the recipe.

The samples where Java programming was involved were tested and run on Java Version 1.7.40. Python Version 2.7 is used wherever Python is used. For MongoDB drivers, you may choose to use the latest available version.

These are pretty common types of software, and their minimum versions are used across different recipes. All the recipes in this book will mention the required software to complete it and their respective version. Some recipes need to be tested on Windows system while some on Linux.

Who this book is for

This book is designed for administrators and developers who are interested in knowing MongoDB and using it as a high-performance and scalable data storage. It is also for those who know the basics of MongoDB and would like to expand their knowledge further. The audience of this book is expected to at least have some basic knowledge of MongoDB.

Sections

In this book, you will find several headings that appear frequently (Getting ready, How to do it, How it works, There's more, and See also).

To give clear instructions on how to complete a recipe, we use these sections as follows:

Getting ready

This section tells you what to expect in the recipe, and describes how to set up any software or any preliminary settings required for the recipe.

How to do it...

This section contains the steps required to follow the recipe.

How it works...

This section usually consists of a detailed explanation of what happened in the previous section.

There's more...

This section consists of additional information about the recipe in order to make the reader more knowledgeable about the recipe.

See also

This section provides helpful links to other useful information for the recipe.

Conventions

In this book, you will find a number of text styles that distinguish between different kinds of information. Here are some examples of these styles and an explanation of their meaning.

Code words in text, database table names, folder names, filenames, file extensions, pathnames, dummy URLs, user input, and Twitter handles are shown as follows: "Create the `/data/mongo/db` directory (or any of your choice)."

Preface

A block of code is set as follows:

```
import com.mongodb.DB;
import com.mongodb.DBCollection;
import com.mongodb.DBObject;
import com.mongodb.MongoClient;
```

Any command-line input or output is written as follows:

```
$ sudo apt-get install default-jdk
```

New terms and **important words** are shown in bold. Words that you see on the screen, for example, in menus or dialog boxes, appear in the text like this: "Without editing any default settings, click on **Launch**."

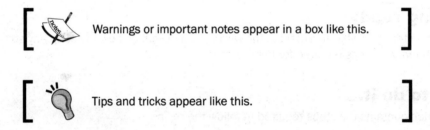

> Warnings or important notes appear in a box like this.

> Tips and tricks appear like this.

Reader feedback

Feedback from our readers is always welcome. Let us know what you think about this book—what you liked or disliked. Reader feedback is important for us as it helps us develop titles that you will really get the most out of.

To send us general feedback, simply e-mail feedback@packtpub.com, and mention the book's title in the subject of your message.

If there is a topic that you have expertise in and you are interested in either writing or contributing to a book, see our author guide at www.packtpub.com/authors.

Customer support

Now that you are the proud owner of a Packt book, we have a number of things to help you to get the most from your purchase.

Downloading the example code

You can download the example code files from your account at http://www.packtpub.com for all the Packt Publishing books you have purchased. If you purchased this book elsewhere, you can visit http://www.packtpub.com/support and register to have the files e-mailed directly to you.

Errata

Although we have taken every care to ensure the accuracy of our content, mistakes do happen. If you find a mistake in one of our books—maybe a mistake in the text or the code—we would be grateful if you could report this to us. By doing so, you can save other readers from frustration and help us improve subsequent versions of this book. If you find any errata, please report them by visiting http://www.packtpub.com/submit-errata, selecting your book, clicking on the **Errata Submission Form** link, and entering the details of your errata. Once your errata are verified, your submission will be accepted and the errata will be uploaded to our website or added to any list of existing errata under the Errata section of that title.

To view the previously submitted errata, go to https://www.packtpub.com/books/content/support and enter the name of the book in the search field. The required information will appear under the **Errata** section.

Piracy

Piracy of copyrighted material on the Internet is an ongoing problem across all media. At Packt, we take the protection of our copyright and licenses very seriously. If you come across any illegal copies of our works in any form on the Internet, please provide us with the location address or website name immediately so that we can pursue a remedy.

Please contact us at copyright@packtpub.com with a link to the suspected pirated material.

We appreciate your help in protecting our authors and our ability to bring you valuable content.

Questions

If you have a problem with any aspect of this book, you can contact us at questions@packtpub.com, and we will do our best to address the problem.

1
Installing and Starting the MongoDB Server

In this chapter, we will cover the following recipes:

- Single node installation of MongoDB
- Starting a single node instance using command-line options
- Single node installation of MongoDB with options from the config file
- Connecting to a single node from the Mongo shell with a preloaded JavaScript
- Connecting to a single node from a Java client
- Starting multiple instances as part of a replica set
- Connecting to the replica set from the shell to query and insert data
- Connecting to the replica set to query and insert data from a Java client
- Starting a simple sharded environment of two shards
- Connecting to a shard from the Mongo shell and performing operations

Introduction

In this chapter, we will look at starting up the MongoDB server. Though it is a cakewalk to start the server for development purposes and with the default settings, there are numerous options that let us tune the startup behavior. We will start the server as a single node; then, we'll introduce various configurations before we conclude by starting up a simple replica set and a sharded setup. So, let's get started by installing and setting up the MongoDB server in the easiest way possible, for simple development purposes.

Installing and Starting the MongoDB Server

Single node installation of MongoDB

In this recipe, we will look at the process of installing MongoDB in the standalone mode. This is the simplest and quickest way to start a MongoDB server but is seldom used for production use cases. However, this is the most common way to start the server for the purpose of development. In this recipe, we will start the server without looking at a lot of other startup options.

Getting ready

Well, assuming that we have downloaded the MongoDB binaries from the download site, extracted them, and have the `bin` directory of MongoDB in the operating system's `path` variable (this is not mandatory but it really becomes convenient), the binaries can be downloaded from http://www.mongodb.org/downloads after selecting your host operating system.

How to do it...

Perform the following steps to start with the single node installation of MongoDB:

1. Create the `/data/mongo/db` directory (or any of your choice). This will be our database directory, and it needs to have permission to let the `mongod` process (the mongo server process) write to it.
2. We will start the server from the console with the `/data/mongo/db` data directory as follows:

   ```
   $ mongod --dbpath   /data/mongo/db
   ```

There's more...

If you see the following message on the console, you have successfully started the server:

```
[initandlisten] waiting for connections on port 27017
```

Starting a server can't get easier than this. Despite the simplicity in starting the server, there are a lot of configuration options that will be used to tune the behavior of the server on startup. Most of the default options are sensible and need not be changed. With the default values, the server should be listening to port `27017` for new connections, and the logs will be printed out to the standard output.

See also

- The *Starting a single node instance using command-line options* recipe for more startup options

Starting a single node instance using command-line options

In this recipe, we will see how to start a standalone single Node server with some command-line options. We will see an example where we will perform the following tasks:

- Starting the server that listens to port `27000`
- Writing logs to `/logs/mongo.log`
- Setting the database directory to `/data/mongo/db`

Since the server is started for development purposes, we don't want to preallocate full size database files (we will soon see what this means).

Getting ready

If you have already seen and executed the steps mentioned in the *Single node installation of MongoDB* recipe, you need not do anything different. If all the prerequisites are met, we are good for this recipe too.

How to do it...

You can start a single node instance using command-line options with the following steps:

1. The `/data/mongo/db` directory for the database and `/logs/` for the logs should be created and present on your filesystem with appropriate write permissions.
2. Execute the following command:

   ```
   > mongod --port 27000 --dbpath /data/mongo/db --logpath /logs/mongo.log --smallfiles
   ```

   Downloading the example code

   You can download the example code files for all Packt books you have purchased from your account at `http://www.packtpub.com`. If you purchased this book elsewhere, you can visit `http://www.packtpub.com/support` and register to have the files e-mailed directly to you.

Installing and Starting the MongoDB Server

How it works...

OK, this wasn't too difficult and is similar to the previous recipe, but we have some additional command-line options this time around. MongoDB actually supports quite a few options at startup, and we will see a list of the ones that are most common and important in my opinion:

Option	Description
`--help` or `-h`	This is used to print the information of various startup options available.
`--config` or `-f`	This specifies the location of the configuration file that contains all the configuration options. We will learn more about this option in the *Single node installation of MongoDB with options from the config file* recipe. It is just a convenient way of specifying the configurations in a file rather than in a command prompt, especially when the number of options specified is more. Using a separate configuration file shared across different `mongod` instances will also ensure that all the instances are running with identical configurations.
`--verbose` or `-v`	This makes the logs more verbose. We can put more v's to make the output even more verbose, for example, `-vvvvv`.
`-quiet`	This is the quieter output. This is the opposite of verbose or the `-v` option. It will keep the logs less chatty and clean.
`--port`	This option is used if you are looking to start the server that listens to a port other than the default `27017`. We will frequently use this option whenever we are looking to start multiple Mongo servers on the same machine; for example, `--port 27018` will start the server that listens to port `27018` for new connections.
`--logpath`	This provides a path to a logfile where the logs will be written. The value defaults to STDOUT. For example, `--logpath /logs/server.out` will use `/logs/server.out` as the logfile for the server. Remember that the value provided should be a file and not a directory where the logs will be written.
`--logappend`	This option will append to the existing logfile if any. The default behavior is to rename the existing logfile and then create a new file for the logs of the currently started Mongo instance. Let's assume that we used the name of the logfile as `server.out` and on startup the file exists. Then, by default, this file will be renamed as `server.out.<timestamp>`, where `<timestamp>` is the current time. The time is GMT as against the local time. Suppose the current date is October 28, 2013 and the time is 12:02:15, then the file generated will have the 2013-10-28T12-02-15 value as the timestamp.
`--dbpath`	This provides the directory where a new database will be created or an existing database is present. The value defaults to `/data/db`. We will start the server using `/data /mongo/db` as the database directory. Note that the value should be a directory rather than the name of the file.

Option	Description
`--smallfiles`	This is used frequently for development purposes when we plan to start more than one Mongo instance on our local machine. On startup, Mongo creates a database file of size 64 MB (on 64-bit machines). This preallocation happens for performance reasons, and the file is created with zeros written to it to fill out the space on the disk. Adding this option on startup creates a preallocated file of 16 MB only (again on a 64-bit machine). This option also reduces the maximum size of the database and journal files. Avoid using this option for production deployments. Also, the database file size doubles to a maximum of 2 GB by default. If the `--smallfile` option is chosen, it goes up to a maximum of 512 MB.
`--replSet`	This option is used to start the server as a member of the replica set. The value of this argument is the name of the replica set, for example, `--replSet repl1`. More information on this option is covered in the *Starting multiple instances as part of a replica set* recipe, where we will start a simple Mongo replica set.
`--configsvr`	This option is used to start the server as a config server. The role of the config server will be made clearer when we set up a simple sharded environment in the *Starting a simple sharded environment of two shards* recipe in this chapter. This, however, will be started and listen to port `27019` by default and the `/data/configdb` data directory. These can, of course, be overridden using the `--port` and `--dbpath` options.
`--shardsvr`	This informs the started `mongod` process that this server is being started as a shard server. By giving this option, the server also listens to port `27018` instead of the default `27017`. We will learn more about this option when we start a simple sharded server.
`--oplogSize`	Oplog is the backbone of replication. It is a capped collection where the data being written to the primary is stored to be replicated to the secondary instances. This collection resides in a database named `local`. On initialization of a replica set, the disk space for the oplog is preallocated, and the database file (for the `local` database) is filled with zeros as placeholders. The default value is 5 percent of the disk space, which should be good enough in most cases. The size of the oplog is crucial, because capped collections are of a fixed size, and they discard the oldest documents in them upon exceeding their size-making space for new documents; if the oplog size is too small, it can result in the data being discarded before being replicated to secondary nodes. A large oplog size can result in unnecessary disk-space utilization and a longer time for the replica set initialization. For development purposes, when we start multiple server processes on the same host, we might want to keep the oplog size to a minimum value so that it quickly initiates the replica set and uses the minimum disk space possible.

Installing and Starting the MongoDB Server

There's more...

For an exhaustive list of the options available, use the `--help` or `-h` option. The preceding list of options is not exhaustive, and we will see some more coming up in the upcoming recipes as and when we need them. In the next recipe, we will see how to use a config file instead of the command-line arguments.

See also

- The *Single node installation of MongoDB with options from the config file* recipe to use config files to provide startup options
- To start a replica set, refer to the *Starting multiple instances as part of a replica set* recipe
- To set up a sharded environment, refer to the *Starting a simple sharded environment of two shards* recipe

Single node installation of MongoDB with options from the config file

As we can see, providing options from the command line does the work, but it starts getting awkward as soon as the number of options we provide increases. We have a nice and clean alternative to providing the startup options from a configuration file rather than as command-line arguments.

Getting ready

If you have already seen and executed the steps mentioned in the *Single node installation of MongoDB* recipe, you need not do anything different, and all the prerequisites of this recipe are the same.

How to do it...

The `/data/mongo/db` directory for the database and `/logs/` for the logs should be created and present on your filesystem, with the appropriate write permissions. Let's take a look at the steps in detail:

1. Create a `config` file that can have any arbitrary name. In our case, let's say we create the file at `/conf/mongo.conf`. We will then edit the file and add the following lines of code to it:

   ```
   port = 27000
   dbpath = /data/mongo/db
   ```

```
logpath = /logs/mongo.log
smallfiles = true
```

2. Start the Mongo server using the following command:

   ```
   > mongod --config  /conf/mongo.conf
   ```

How it works...

All the command-line options we discussed in the previous recipe, *Starting a single node instance using command-line options*, hold true. We are just providing these options in a configuration file instead. If you have not visited the previous recipe, I recommend that you do so, as this is where we have discussed some of the common command-line options. The properties are specified as `<property name> = <value>`. For all those properties that don't have values, for example, the `smallfiles` option, the value given is a Boolean value, `true`. If you need to have a verbose output, you will add `v=true` (or multiple v's to make it more verbose) to our `config` file. If you already know what the command-line option is, it is pretty easy to guess the value of the property in the file. It is the similar to the command-line option, with just the hyphen removed.

Connecting to a single node from the Mongo shell with a preloaded JavaScript

This recipe is about starting the Mongo shell and connecting to a MongoDB server. Here, we'll also demonstrate how to load JavaScript code into the shell. Though this is not always required, it is handy when we have a large block of JavaScript code, including variables and functions with some business logic in them that is required to be executed from the shell frequently, and we want these functions to be available in the shell always.

Getting ready

It is not necessary for the MongoDB server to run to start a shell. We will rarely start a shell without connecting it to a running MongoDB server. To start a server on the localhost without much of a hassle, take a look at the first recipe, *Single node installation of MongoDB*, and start the server.

How to do it...

Let's take a look at the steps in detail:

1. First, we will start by creating a simple JavaScript file; let's call it `hello.js`. Type in the following lines in the `hello.js` file:

   ```
   function sayHello(name) {
   ```

Installing and Starting the MongoDB Server

```
   print('Hello ' + name + ', how are you?')
}
```

2. Save this file at `/mongo/scripts`. (it can be saved at any other location too).
3. In the command prompt, execute the following command:

   `> mongo --shell /mongo/scripts/hello.js`

4. On executing this, we should see the following message on our console:

   `MongoDB shell version: 2.4.6`
   `connecting to: test`
   `>`

5. Test the database that the shell is connected to by typing the following command:

   `> db`

   This should print out `test` on the console.

6. Now, type in the following command on the shell:

   `> sayHello('Fred')`
   `Hello Fred, how are you?`

How it works...

The JavaScript function we executed here is of no practical use, but it's just used to demonstrate how a function can be preloaded upon the startup of the shell. There can be multiple functions in the `.js` file that contain valid JavaScript code, possibly some complex business logic.

When we executed the `mongo` command without any arguments, we connected to the MongoDB server that runs on the localhost and listens for new connections on the default port `27017`. The format of the command is as follows:

`mongo <options> <db address> <.js files>`

If there are no arguments passed to the `mongo` executable, it is equivalent to passing db address as `localhost:27017/test`.

There's more...

Let's look at some example values of the `db address` command-line option and its interpretation:

- `mydb`: This will connect to the server that runs on the localhost and listens for connection on port `27017`. The database connected will be `mydb`.
- `mongo.server.host/mydb`: This will connect to the server that runs on `mongo.server.host` and the default port `27017`. The database connected will be `mydb`.

- `mongo.server.host:27000/mydb`: This will connect to the server that runs on `mongo.server.host` and the port `27000`. The database connected will be `mydb`.
- `mongo.server.host:27000`: This will connect to the server that runs on `mongo.server.host` and the port `27000`. The database connected will be the default database, `test`.

Now, there are quite a few options available on the Mongo client too. We will see a few of them in the following table:

Option	Description
`--help` or `-h`	This offers help regarding the usage of various command-line options.
`--shell`	When `.js` files are given as arguments, these scripts get executed, and the Mongo client will exit. Providing this option ensures that the shell remains running after the JavaScript files execute. All the functions and variables defined in these `.js` files are available in the shell upon startup. As in the preceding case, the `sayHello` function defined in the JavaScript file is available in the shell for invocation.
`--port`	This specifies the port of the Mongo server where the client needs to connect.
`--host`	This specifies the hostname of the Mongo server where the client needs to connect. If the db address is provided with the hostname, port, and database, both the `--host` and `--port` options need not be specified.
`--username` or `-u`	This is relevant when security is enabled for Mongo. It is used to provide the username of the user to be logged in.
`--password` or `-p`	This is relevant when security is enabled for Mongo. It is used to provide the password of the user to be logged in.

Connecting to a single node from a Java client

This recipe is about setting up the Java client for MongoDB. You will be repeatedly referring to this recipe while working on others, so read it very carefully.

Getting ready

The following are the prerequisites for this recipe:

- Version 1.6 or above of Java SDK is recommended.
- Use the latest available version of Maven. Version 3.1.1 was the latest at the time of writing this book.

Installing and Starting the MongoDB Server

- Use the MongoDB Java driver. Version 2.11.3 was the latest at the time of writing this book.
- Connectivity to the Internet to access the online Maven repository or a local repository is needed. Alternatively, you might choose an appropriate local repository accessible to you from your computer.
- The Mongo server is up and running on the localhost and on port `27017`. Take a look at the first recipe, *Single node installation of MongoDB*, and start the server.

How to do it...

Let's take a look at the steps in detail:

1. Install the latest version of JDK if you don't already have it on your machine. We will not be going through the steps to install JDK in this recipe but, before moving on with, next step, the JDK should be present. Type `javac -version` on the shell to check for the version installed.

2. Once the JDK is set up, the next step is to set up Maven. Skip the next three steps if Maven is already installed on your machine.

3. Maven needs to be downloaded from `http://maven.apache.org/download.cgi`. Choose the binaries in the `.tar.gz` or `.zip` format and download it. This recipe is executed on a machine that runs on the Windows platform; thus, these steps are for installation on Windows. The following screenshot shows the download page of Maven:

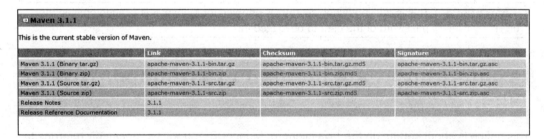

4. Once the archive is downloaded, we need to extract it and put the absolute path of the `bin` folder in the extracted archive in the operating system's `path` variable. Maven also needs the path of the JDK to be set as the `JAVA_HOME` environment variable. Remember to set the root of your JDK as the value of this variable.

5. All we need to do now is type `mvn -version` in the command prompt. If you see the version of Maven on the command prompt, we have successfully set up Maven:

    ```
    > mvn -version
    ```

6. At this stage, we have Maven installed, and we are now ready to create our simple project to write our first Mongo client in Java. We will start by creating a `project` folder. Let's assume that we create a folder called `Mongo Java`. Then, we will create a folder structure `src/main/java` in this `project` folder. The root of the project folder then contains a file called `pom.xml`. Once this folder creation is done, the folder structure should look as follows:

    Mongo Java

    +–src

    | +main

    | +java

    |–pom.xml

7. We just have the project skeleton with us now. We will now add some content to the `pom.xml` file. Not much is needed for this. Add the following code snippet in the `pom.xml` file and save it:

    ```xml
    <project>
        <modelVersion>4.0.0</modelVersion>
        <name>Mongo Java</name>
        <groupId>com.packtpub</groupId>
        <artifactId>mongo-cookbook-java</artifactId>
        <version>1.0</version>
        <packaging>jar</packaging>
        <dependencies>
            <dependency>
                <groupId>org.mongodb</groupId>
                <artifactId>mongo-java-driver</artifactId>
                <version>2.11.3</version>
            </dependency>
        </dependencies>
    </project>
    ```

8. Finally, we will write our Java client that will be used to connect to the Mongo server and execute some very basic operations. The following is the Java class located at `src/main/java` in the `com.packtpub.mongo.cookbook` package, and the name of the class is `FirstMongoClient`:

    ```java
    package com.packtpub.mongo.cookbook;

    import com.mongodb.BasicDBObject;
    import com.mongodb.DB;
    import com.mongodb.DBCollection;
    import com.mongodb.DBObject;
    ```

Installing and Starting the MongoDB Server

```java
import com.mongodb.MongoClient;

import java.net.UnknownHostException;
import java.util.List;

/**
 * Simple Mongo Java client
 *
 */
public class FirstMongoClient {

    /**
     * Main method for the First Mongo Client. Here we shall be connecting to a mongo
     * instance running on localhost and port 27017.
     *
     * @param args
     */
    public static final void main(String[] args) throws UnknownHostException {
        MongoClient client = new MongoClient("localhost", 27017);
        DB testDB = client.getDB("test");
        System.out.println("Dropping person collection in test database");
        DBCollection collection = testDB.getCollection("person");
        collection.drop();
        System.out.println("Adding a person document in the person collection of test database");
        DBObject person =
new BasicDBObject("name", "Fred").append("age", 30);
        collection.insert(person);
        System.out.println("Now finding a person using findOne");
        person = collection.findOne();
        if(person != null) {
            System.out.printf("Person found, name is %s and age is %d\n", person.get("name"), person.get("age"));
        }
        List<String> databases = client.getDatabaseNames();
        System.out.println("Database names are");
        int i = 1;
        for(String database : databases) {
            System.out.println(i++ + ": " + database);
        }
```

```
    System.out.println("Closing client");
        client.close();
  }
}
```

9. It's now time to execute the preceding Java code. We will execute it using Maven from the shell. You should be in the same directory as the `pom.xml` file of the project:

   mvn compile exec:java -Dexec.mainClass=com.packtpub.mongo.cookbook.FirstMongoClient

How it works...

Those were quite a lot of steps to follow! Let's look at some of them in more detail. Everything up to step 6 is straightforward and doesn't need any explanation. Let's look at the other steps.

The `pom.xml` file we have here is pretty simple. We defined a dependency on Mongo's Java driver. It relies on the online repository (http://search.maven.org) for resolving the artifacts. For a local repository, all we need to do is define the repositories and `pluginRepositories` tags in `pom.xml`. For more information on Maven, refer to the Maven documentation at http://maven.apache.org/guides/index.html.

Now, for the Java class, the `org.mongodb.MongoClient` class is the backbone. We will first instantiate it using one of its overloaded constructors that gives the server's host and port. In this case, the hostname and port were not really needed as the values provided are the default values anyway, and the no-argument constructor would have worked well too. The following line of code instantiates this client:

```
MongoClient client = new MongoClient("localhost", 27017);
```

The next step is to get the database; in this case, `test` using the `getDB` method. This is returned as an object of type `com.mongodb.DB`. Note that this database might not exist, yet `getDB` will not throw any exception. Instead, the database will get created whenever we add a new document to the collection in this database. Similarly, `getCollection` on the `DB` object will return an object of type `com.mongodb.DBCollection`, representing the collection in the database. This too might not exist in the database and will get created automatically upon the insertion of the first document.

The following lines of code from our class show how to get an instance of `DB` and `DBCollection`:

```
DB testDB = client.getDB("test");
DBCollection collection = testDB.getCollection("person");
```

Installing and Starting the MongoDB Server

Before we insert a document, we will drop the collection so that even upon multiple executions of the program, we will have just one document in the `person` collection. The collection is dropped using the `drop()` method on the `DBCollection` object's instance. Next, we will create an instance of `com.mongodb.DBObject`. This is an object that represents the document to be inserted in the collection. The concrete class used here is `BasicDBObject`, which is a type of `java.util.LinkedHashMap` class, where the key is a string and the value is an object. The value can be another `DBObject`, too, in which case it is a document nested within another document. In our case, we have two keys: `name` and `age`. These are the field names in the document to be inserted, and the values are of type string and integer, respectively. The append method of `BasicDBObject` adds a new key-value pair to the `BasicDBObject` instance and returns the same instance, which allows us to chain the `append` method calls to add multiple key value pairs. `DBObject` is then inserted into the collection using the `insert` method. This is how we instantiated a `DBObject` for the person and inserted it in the collection:

```
DBObject person = new BasicDBObject("name", "Fred").append("age", 30);
collection.insert(person);
```

The `findOne` method on `DBCollection` is straightforward and returns one document from the collection. This version of `findOne` doesn't accept `DBObject` (which, otherwise, acts as a query executed before a document is selected and returned) as a parameter. This is synonymous to executing a `db.person.findOne()` from the Mongo shell.

Finally, we will simply invoke `getDatabaseNames` to get a list of databases names in the server. At this point of time, we should at least be having the `test` and `local` databases in the returned result. Once all the operations are completed, we will close the client. The `MongoClient` class is thread-safe; generally, one instance is used per application. To execute the program, we will use Maven's exec plugin. On executing step 9, we will see the following output on the console:

```
[INFO] --- exec-maven-plugin:1.2.1:java (default-cli) @ mongo-cookbook-java ---
Dropping person collection in test database
Adding a person document in the person collection of test database
Now finding a person using findOne
Person found, name is Fred and age is 30
Database names are
1: local
2: test
Closing client
[INFO] ------------------------------------------------------------------------
```

```
[INFO] BUILD SUCCESS
[INFO] ------------------------------------------------------------------------
[INFO] Total time: 5.183s
[INFO] Finished at: Wed Oct 30 00:42:29 IST 2013
[INFO] Final Memory: 7M/19M
[INFO] ------------------------------------------------------------------------
```

Starting multiple instances as part of a replica set

In this recipe, we will look at starting multiple servers on the same host but as a cluster. Starting a single Mongo server is enough for development purposes or applications that are not mission-critical. For crucial production deployments, we need the availability to be high where, if one server instance fails, another instance takes over and the data remains available for querying, inserting, or updating. Clustering is an advanced concept, and we won't be doing it justice by covering this whole concept in one recipe. In this recipe, we will touch the surface and get into more details in other recipes in *Chapter 4, Administration*, later in the book. In this recipe, we will start multiple Mongo server processes on the same machine for testing purpose. In the production environment, they will be running on different machines (or virtual machines) in the same or different data centers.

Let's see in brief exactly what a replica set is. As the name suggests, it is a set of servers that are replicas of each other in terms of data. Looking at how they are kept in sync with each other and other internals is something we will defer to some later recipes in *Chapter 4, Administration*, but one thing to remember is that write operations will happen only on one node, the primary one. All the querying also happens from the primary node by default, though we might permit read operations on secondary instances explicitly. An important fact to remember is that replica sets are not meant to achieve scalability by distributing the read operations across various nodes in a replica set. Their sole objective is to ensure high availability.

Getting ready

Though not a prerequisite, taking a look at the *Starting a single node instance using command-line options* recipe will definitely make things easier, just in case you are not aware of the various command-line options and their significance while starting a Mongo server. Also, the necessary binaries and setup as mentioned in the *Single node installation of MongoDB* recipe must be mastered before we continue with this recipe. Let's sum up what we need to do.

Installing and Starting the MongoDB Server

We will start three mongod processes (Mongo server instances) on our localhost. Then, we will create three data directories, `/data/n1`, `/data/n2`, and `/data/n3`, for node 1, node 2, and node 3, respectively. Similarly, we will redirect the logs to `/logs/n1.log`, `/logs/n2.log`, and `/logs/n3.log`. The following diagram will give you an idea as to how the cluster will look like:

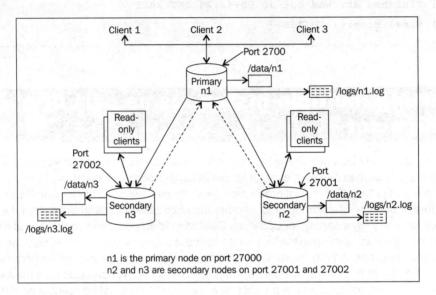

How to do it...

Let's take a look at the steps in detail:

1. Create the `/data/n1`, `/data/n2`, and `/data/n3` directories, `/logs` for data, and logs of the three nodes. On the Windows platform, you can choose the `c:\data\n1`, `c:\data\n2`, `c:\data\n3`, or `c:\logs\` directory (or any other directory of your choice) for data and logs, respectively. Ensure that these directories have appropriate write permissions for the Mongo server to write the data and logs.

2. Start the three servers as follows (note that users on the Windows platform need to skip the `--fork` option, as it is not supported):

   ```
   $ mongod --replSet repSetTest --dbpath /data/n1 --logpath /logs/n1.log --port 27000 --smallfiles --oplogSize 128 --fork
   $ mongod --replSet repSetTest --dbpath /data/n2 --logpath /logs/n2.log --port 27001 --smallfiles --oplogSize 128 --fork
   $ mongod --replSet repSetTest --dbpath /data/n3 --logpath /logs/n3.log --port 27002 --smallfiles --oplogSize 128 --fork
   ```

3. Start the Mongo shell and connect to any of the Mongo servers that are running. In this case, we will connect to the first one (the one listening to port 27000). Execute the following command:

   ```
   $ mongo localhost:27000
   ```

4. Try to execute an insert operation from the Mongo shell after connecting to it as follows:

   ```
   > db.person.insert({name:'Fred', age:35})
   ```

   This operation should fail as the replica set is not initialized yet. More information can be found in the *How it works...* section of this recipe.

5. The next step is to start configuring the replica set. We will start by preparing a JSON configuration in the shell:

   ```
   cfg = {
       '_id':'repSetTest',
       'members':[
           {'_id':0, 'host': 'localhost:27000'},
           {'_id':1, 'host': 'localhost:27001'},
           {'_id':2, 'host': 'localhost:27002'}
       ]
   }
   ```

6. The last step is to initiate the replica set with the preceding configuration as follows:

   ```
   > rs.initiate(cfg)
   ```

Execute rs.status() after a few seconds on the shell to see the status. In a few seconds, one of them should become primary, and the remaining two should become secondary.

How it works...

We described the common options and all these command-line options in the *Starting a single node instance using command-line options* recipe in detail.

As we are starting three independent mongod services, we have three dedicated database paths on the filesystem. Similarly, we have three separate logfile locations for each of the processes. We then started three mongod processes with the database and logfile path specified. As this setup is for test purposes and started on the same machine, we used the --smallfiles and --oplogSize options. Avoid using these options in the production environment. As these are running on the same host, we also choose the ports explicitly to avoid port conflicts. The ports we chose here are 27000, 27001, and 27002. When we start the servers on different hosts, we might or might not choose a separate port. We can very well choose to use the default one whenever possible.

Installing and Starting the MongoDB Server

The `--fork` option demands some explanation. By choosing this option, we started the server as a background process from our operating system's shell and got the control back in the shell, where we can then start more such `mongod` processes or perform other operations. In the absence of the `--fork` option, we cannot start more than one process per shell and will need to start three `mongod` processes in three separate shells. This option, however, doesn't work on the Windows platform, and we need to start one process per shell. We can, however, execute the following command to spawn a new shell and then start the new Mongo service in this newly spawned shell:

```
start mongod --replSet repSetTest --dbpath c:\data\c1 --logpath c:\logs\
n1.log --port 27000 --smallfiles --oplogSize 128
```

The preceding command allows us to have a batch file (a `.bat` file) that contains all the logic to create the relevant directories and then spawn three `mongod` processes in three shells.

Let's get back to the replica set creation; we are not yet done with setting up a replica set. If we take a look at the logs generated in the `log` directory, we will see the following lines in it:

```
[rsStart] replSet can't get local.system.replset config from self or
any seed (EMPTYCONFIG)
[rsStart] replSet info you may need to run replSetInitiate --
rs.initiate() in the shell -- if that is not already done
```

Though we started three `mongod` processes with the `--replSet` option, we still haven't configured them to work with each other as a replica set. This command-line option is just used to tell the server on startup that this process will be running as part of a replica set. The name of the replica set is the same as the value of this option passed on the command prompt. This also explains why the `insert` operation executed on one of the nodes failed before the replica set was initialized. In mongo replica sets, only one node is the primary node where all the inserts and querying happen. In the preceding diagram, node n1 is shown as the primary node and listens to port `27000` for client connections. All the other nodes are slave/secondary instances that sync themselves up with the primary node; hence, querying too is disabled on them by default. It is only when the primary node goes down that one of the secondaries takes over and becomes a primary node. It is, however, possible to query the secondary instances for data, as we showed in the preceding diagram. We will see how to query from a secondary instance in the next recipe.

Well, all that is left now is to configure the replica set by grouping the three processes we started. This is done by first defining a JSON object as follows:

```
cfg = {
  '_id':'repSetTest',
  'members':[
        {'_id':0, 'host': 'localhost:27000'},
        {'_id':1, 'host': 'localhost:27001'},
        {'_id':2, 'host': 'localhost:27002'}
   ]
}
```

There are two fields, _id and members, for the unique ID of the replica set and an array of the hostnames and port numbers of the mongod server processes as part of this replica set, respectively. Using the localhost to refer to the host is not a very good idea and is usually discouraged. However, in this case, we started all the processes on the same machine; thus, we are OK with it. It is, however, preferred to refer to the hosts by their hostnames even if they are running on the localhost. Note that you cannot mix referring the instances using the localhost and hostnames both in the same config. You can use either the hostnames or the localhost. To configure the replica set, we then connect to any one of three running mongod processes; in this case, we will connect to the first one and then execute the following command from the shell:

```
> rs.initiate(cfg)
```

The _id in the cfg object passed has the same value as the value we gave to the --replSet option in the command prompt when we started the server processes. Not giving the same value will throw the following error:

```
{
        "ok" : 0,
        "errmsg" : "couldn't initiate : set name does not match the set name host Amol-PC:27000 expects"
}
```

If all goes well and the initiate call is successful, you will see something like the following JSON response on the shell:

```
{
        "info" : "Config now saved locally.  Should come online in about a minute.",
        "ok" : 1
}
```

In a few seconds, you should see a different prompt for the shell from which we executed this command. It should now become a primary or secondary node. The following command is an example of the shell connected to a primary member of the replica set:

repSetTest:PRIMARY>

Executing rs.status() should give us some stats on the replica set status. The stateStr field here is important, and it contains the text PRIMARY, SECONDARY, and so on.

Installing and Starting the MongoDB Server

There's more...

If you are looking to convert a standalone instance to a replica set, the instance with data needs to become a primary instance first, and then empty secondary instances will be added, to which the data will be synchronized. For more information on how to perform this operation, visit http://docs.mongodb.org/manual/tutorial/convert-standalone-to-replica-set/.

See also

- The *Connecting to the replica set from the shell to query and insert data* recipe to perform more operations from the shell after connecting to a replica set
- *Chapter 4, Administration*, for more advanced recipes on replication

Connecting to the replica set from the shell to query and insert data

In the previous recipe, we started a replica set of three mongod processes. In this recipe, we will be working on top of it and will connect to it from the client application, perform querying, insert data, and take a look at some of the interesting aspects of the replica set from a client's perspective.

Getting ready

The prerequisite for this recipe is that the replica set should be set up, and it should be up and running. For details on how to start the replica set, refer to the *Starting multiple instances as part of a replica set* recipe.

How to do it...

Let's take a look at the steps in detail:

1. Create the /data/n1, /data/n2, /data/n3, and /logs directories for data and logs of the three nodes, respectively.
2. We will start two shells here: one for primary and one for secondary. Execute the following command in the command prompt:

   ```
   mongo localhost:27000
   ```

3. The prompt of the shell tells whether the server to which we connected is primary or secondary. It should show the replica set's name followed by : and then followed by the server's state. In this case, if the replica set is initialized and is up and running, we will see either `repSetTest:PRIMARY>` or `repSetTest:SECONDARY>`.

4. Suppose the first server we connected to is a secondary server, then we need to find the primary server as follows:
    1. Execute the `rs.status()` command in the shell and look out for the `stateStr` field. This should give us the primary server. Use the Mongo shell to connect to this server. At this point, we should have two shells running: one connected to a primary node and the other connected to a secondary node.

5. In the shell connected to the primary node, execute the following insert command:

   `repSetTest:PRIMARY> db.replTest.insert({_id:1, value:'abc'})`

   There is nothing special about it. We have just inserted a small document in a collection that we use for the replication test.

6. By executing the following query on the primary node, we should get one result:

   `repSetTest:PRIMARY> db.replTest.findOne()`

   `{ "_id" : 1, "value" : "abc" }`

7. So far so good. Now, we will go to the shell that is connected to the secondary node and execute the following command:

   `repSetTest:SECONDARY> db.replTest.findOne()`

8. On doing this, we will see the following error on the console:

   `{ "$err" : "not master and slaveOk=false", "code" : 13435 }`

9. Now, execute the following command on the console:

   `repSetTest:SECONDARY>  rs.slaveOk(true)`

10. Execute the query we executed in step 7 again on the shell. This will now get the following results:

    `repSetTest:SECONDARY>db.replTest.findOne()`

    `{ "_id" : 1, "value" : "abc" }`

11. Execute the following insert command on the secondary node; it should not succeed with the following message:

    `repSetTest:SECONDARY> db.replTest.insert({_id:1, value:'abc'}) not master`

How it works...

We have done a lot of things in this recipe, and we will try to throw some light on some of the important concepts to remember.

We basically connected to a primary and a secondary node from the shell and performed (I would say, tried to perform) the select and insert operations. The architecture of a Mongo replica set is made up of one primary (just one; no more, no less) and multiple secondary nodes. All writes happen on the primary node only. Note that replication is not a mechanism to distribute a read-request load that enables us to scale the system. Its primary intent is to ensure high availability of data. By default we are not permitted to read data from the secondary nodes. In step 6, we simply inserted data from the primary node and then executed the query to get the document that we inserted. This is straightforward, and there is nothing related to clustering here. Just note that we inserted the document from the primary node and then queried it back.

In the next step, we executed the same query but, this time, from the secondary node's shell. By default, querying is not enabled on the secondary node. There might be a small lag in replicating the data, possibly due to heavy data volumes to be replicated, network latency, and hardware capacity to name a few of the causes; thus, querying on the secondary node might not reflect the latest inserts or updates made on the primary node. If, however, we are OK with it and can live with the slight lag in the data being replicated, all we need to do is enable querying on the secondary node explicitly by just executing one command, `rs.slaveOk()` or `rs.slaveOk(true)`. Once this is done, we are free to execute queries on the secondary nodes too.

Finally, we tried to insert data in a collection of the slave node. Under no circumstances this is permitted, regardless of whether we have executed `rs.slaveOk()`. When `rs.slaveOk()` is invoked, it just permits the data to be queried from the secondary node. All the write operations still have to go to the primary node and then flow down to the secondary node. The internals of replication will be covered in a different recipe in the *Understanding and analyzing oplogs* recipe in *Chapter 4, Administration*.

See also

- The *Connecting to the replica set to query and insert data from a Java client* recipe is to get details on how to connect to replica set from a Java client

Connecting to the replica set to query and insert data from a Java client

In this recipe, we will demonstrate how to connect to a replica set using a Java client and execute queries and insert data using the Java client for MongoDB. We will also see how the client would automatically failover to another member in the replica set should a primary member goes down.

Getting ready

We first need to take a look at the *Connecting to a single node from a Java client* recipe, as it contains all the prerequisites and steps to set up Maven and other dependencies. As we are dealing with a Java client for replica sets, a replica set must be up and running. Refer to the *Starting multiple instances as part of a replica set* recipe for details on how to start the replica set.

How to do it...

Let's take a look at the steps in detail:

1. First, we need to write/copy the following piece of code (this Java class is also available for download from the book's site):

    ```java
    package com.packtpub.mongo.cookbook;

    import com.mongodb.BasicDBObject;
    import com.mongodb.DB;
    import com.mongodb.DBCollection;
    import com.mongodb.DBObject;
    import com.mongodb.MongoClient;
    import com.mongodb.ServerAddress;

    import java.util.Arrays;

    /**
     *
     */
    public class ReplicaSetMongoClient {

        /**
         * Main method for the test client connecting to the replica set.
         * @param args
    ```

```java
     */
    public static final void main(String[] args) throws Exception {
        MongoClient client = new MongoClient(
                Arrays.asList(
                        new ServerAddress("localhost", 27000),
                        new ServerAddress("localhost", 27001),
                        new ServerAddress("localhost", 27002)
                )
        );
        DB testDB = client.getDB("test");
        System.out.println("Dropping replTest collection");
        DBCollection collection = testDB.getCollection("replTest");
        collection.drop();
        DBObject object = new BasicDBObject("_id", 1).append("value", "abc");
        System.out.println("Adding a test document to replica set");
        collection.insert(object);
        System.out.println("Retrieving document from the collection, this one comes from primary node");
        DBObject doc = collection.findOne();
        showDocumentDetails(doc);
        System.out.println("Now Retrieving documents in a loop from the collection.");
        System.out.println("Stop the primary instance manually after few iterations");
        for(int i = 0 ; i < 20; i++) {
            try {
                doc = collection.findOne();
                showDocumentDetails(doc);
            } catch (Exception e) {
                //Ignoring or log a message
            }
            Thread.sleep(5000);
        }
    }

    /**
     *
     * @param obj
     */
    private static void showDocumentDetails(DBObject obj) {
        System.out.printf("_id: %d, value is %s\n", obj.get("_id"), obj.get("value"));
    }
}
```

Chapter 1

2. Connect to any of the nodes in the replica set, say to `localhost:27000`, and, from the shell, execute `rs.status()`. Take a note of the primary instance in the replica set and connect to it from the shell if `localhost:27000` is not a primary node. Now, switch to the admin database as follows:

   repSetTest:PRIMARY>use admin

3. Now, execute the preceding program from the operating system shell as follows:

   $ mvn compile exec:java -Dexec.mainClass=com.packtpub.mongo. cookbook.ReplicaSetMongoClient

4. Shut down the primary instance by executing the following command on the Mongo shell connected to the primary node:

   repSetTest:PRIMARY> db.shutdownServer()

5. Watch the output on the console where the `com.packtpub.mongo.cookbook. ReplicaSetMongoClient` class is executed using Maven.

How it works...

An interesting thing to observe is how we instantiate a `MongoClient` instance. It is done as follows:

```
MongoClient client = new MongoClient(Arrays.asList(
   new ServerAddress("localhost", 27000),
   new ServerAddress("localhost", 27001),
   new ServerAddress("localhost", 27002)));
```

The constructor takes a list of `com.mongodb.ServerAddress`. This class has a lot of overloaded constructors, but we chose to use the one that takes the hostname and port number. Here, we provided all the server details in a replica set as a list. We haven't mentioned what the primary node is and what the secondary nodes are. The `MongoClient` class is intelligent enough to figure this out and connect to the appropriate instance. The list of servers provided is called the seed list. It need not contain an entire set of servers in a replica set, though the objective is to provide as much as we can. The `MongoClient` class will figure out all the server details from the provided subset. For example, if the replica set is of five nodes but we provide only three servers, it still works fine. On connecting with the provided replica set servers, the client will query them to get the replica set metadata and figure out the rest of the provided servers in the replica set. In the preceding case, we instantiated the client with three instances in the replica set. If the replica set has five members, instantiating the client with just three of them as we did earlier is still good enough, and the remaining two instances will be automatically discovered.

Installing and Starting the MongoDB Server

Next, we will start the client from the command prompt using Maven. Once the client is running in the loop to find one document, we will bring down the primary instance. We will see something like the following output on the console:

```
_id: 1, value is abc
Now, retrieving documents in a loop from the collection.
Stop the primary instance manually after a few iterations:
_id: 1, value is abc
_id: 1, value is abc
Nov 03, 2013 5:21:57 PM com.mongodb.ConnectionStatus$UpdatableNode update
WARNING: Server seen down: Amol-PC/192.168.1.171:27002
java.net.SocketException: Software caused connection abort: recv failed
        at java.net.SocketInputStream.socketRead0(Native Method)
        at java.net.SocketInputStream.read(SocketInputStream.java:150)
    …
WARNING: Primary switching from Amol-PC/192.168.1.171:27002 to Amol-PC/192.168.1.171:27001
_id: 1, value is abc
```

As we can see, the query in the loop was interrupted when the primary node went down. The client, however, switched to the new primary node seamlessly, well, nearly seamlessly, as the client might have to catch an exception and retry the operation after a predetermined interval has elapsed.

Starting a simple sharded environment of two shards

In this recipe, we will set up a simple sharded setup made up of two data shards. There will be no replication to keep it simple, as this is the most basic shard setup to demonstrate the concept. We won't be getting deep into the internals of sharding, which we will explore further in *Chapter 4, Administration*.

Here is a bit of theory before we proceed. Scalability and availability are two important cornerstones for building any mission-critical application. Availability is something that was taken care of by replica sets, which we discussed in the previous recipes of this chapter. Let's look at scalability now. Simply put, scalability is the ease with which the system can cope with an increasing data and request load. Consider an e-commerce platform. On regular days, the number of hits to the site and load is fairly modest, and the system response times and error rates are minimal (this is subjective).

Now, consider the days where the system load becomes twice or three times an average day's load (or even more), for example, say on Thanksgiving Day, Christmas, and so on. If the platform is able to deliver similar levels of service on these high-load days compared with any other day, the system is said to have scaled up well to the sudden increase in the number of requests.

Now, consider an archiving application that needs to store the details of all the requests that hit a particular website over the past decade. For each request that hits the website, we will create a new record in the underlying data store. Suppose each record is of 250 bytes with an average load of 3 million requests per day, then we will cross the 1 TB data mark in about 5 years. This data will be used for various analytic purposes and might be frequently queried. The query performance should not be drastically affected when the data size increases. If the system is able to cope with this increasing data volume and still gives a decent performance comparable to that on low data volumes, the system is said to have scaled up well against the increasing data volumes.

Now that we have seen in brief what scalability is, let me tell you that sharding is a mechanism that lets a system scale to increasing demands. The crux lies in the fact that the entire data is partitioned into smaller segments and distributed across various nodes called shards. Let's assume that we have a total of 10 million documents in a Mongo collection. If we shard this collection across 10 shards, we will ideally have 10,000,000/10 = 1,000,000 documents on each shard. At a given point of time, one document will only reside on one shard (which, by itself, will be a replica set in a production system). There is, however, some magic involved that keeps this concept hidden from the developer querying the collection, who gets one unified view of the collection irrespective of the number of shards. Based on the query, it is Mongo that decides which shard to query for the data and return the entire result set. With this background, let's set up a simple shard and take a closer look at it.

Getting ready

Apart from the MongoDB server already installed, there are no prerequisites from a software perspective. We will create two data directories, one for each shard. There will be one directory for data and one for logs.

How to do it...

Let's take a look at the steps in detail:

1. We will start by creating directories for logs and data. Create the `/data/s1/db`, `/data/s2/db`, and `/logs` directories. On Windows, we can have `c:\data\s1\db`, and so on for the `data` and `log` directories. There is also a config server that is used in a sharded environment to store some metadata. We will use `/data/con1/db` as the data directory for the config server.

Installing and Starting the MongoDB Server

2. Start the following mongod processes, one for each of the two shards and one for the config database, and one mongos process (we will see what this process does). For the Windows platform, skip the --fork parameter as it is not supported:

   ```
   $ mongod --shardsvr --dbpath /data/s1/db --port 27000 --logpath /logs/s1.log --smallfiles --oplogSize 128 --fork
   $ mongod --shardsvr --dbpath /data/s2/db --port 27001 --logpath /logs/s2.log --smallfiles --oplogSize 128 --fork
   $ mongod --configsvr --dbpath /data/con1/db --port 25000 --logpath /logs/config.log --fork
   $ mongos --configdb localhost:25000 --logpath /logs/mongos.log --fork
   ```

3. In the command prompt, execute the following command. This will show a mongos prompt:

   ```
   $ mongo
   MongoDB shell version: 2.4.6
   connecting to: test
   mongos>
   ```

4. Finally, we set up the shard. From the mongos shell, execute the following two commands:

   ```
   mongos> sh.addShard("localhost:27000")
   mongos> sh.addShard("localhost:27001")
   ```

5. On the addition of each shard, we will get an ok reply. Something like the following JSON message will be seen giving the unique ID for each shard that is added:

   ```
   { "shardAdded" : "shard0000", "ok" : 1 }
   ```

> We have used localhost everywhere to refer to the locally running servers. It is not a recommended approach and is discouraged. A better approach will be to use hostnames even if they are local processes.

How it works

Let's see what we did in the process. We created three directories for data (two for the shards and one for the config database) and one directory for logs. We can have a shell script or a batch file to create the directories as well. In fact, in large production deployments, setting up shards manually is not only time-consuming but also error-prone.

Let's try to get a picture of what exactly we have done and what we are trying to achieve.

The following diagram shows the shard setup we just built:

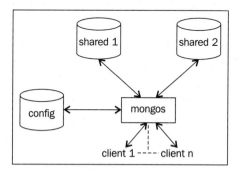

If we look at the preceding diagram and the servers started in step 2, we will see that we have shard servers that will store the actual data in the collections. These were the first two of the four processes that started listening to port `27000` and `27001`. Next, we started a config server, which is seen on the left-hand side in the preceding diagram. It is the third server of the four servers started in step 2, and it listens to port `25000` for incoming connections. The sole purpose of this database is to maintain the metadata of the shard servers. Ideally, only the `mongos` process or drivers connect to this server for the shard details/metadata and the shard key information. We will see what a shard key is in the next recipe, where we will play around with a sharded collection and see the shards we created in action.

Finally, we have a `mongos` process. This is a lightweight process that doesn't do any persistence of data and just accepts connections from clients. This is the layer that acts as a gatekeeper and abstracts the client from the concept of shards. For now, we can view it as a router that consults the config server and takes the decision to route the client's query to the appropriate shard server for execution. It then aggregates the result from various shards if applicable and returns the result to the client. It is safe to say that no client directly connects to the config or the shard servers; in fact, ideally, no one should connect to these processes directly, except for some administration operations. Clients simply connect to the `mongos` process and execute their queries, or insert or update operations.

Installing and Starting the MongoDB Server

Just by starting the shard servers, the config server and `mongos` process don't create a sharded environment. On starting up the `mongos` process, we provided it with the details of the config server. What about the two shards that will be storing the actual data? The two `mongod` processes that started as shard servers are, however, not yet declared anywhere as shard servers in the configuration. That is exactly what we do in the final step by invoking `sh.addShard()` for both the shard servers. The `mongos` process is provided with the config server's details on startup. Adding shards from the shell stores this metadata about the shards in the config database; then, the `mongos` processes will query this config database for the shard's information. On executing all the steps of this recipe, we will have an operational shard. Before we conclude, the shard we set up here is far from ideal and not how it will be done in the production environment. The following diagram gives us an idea of how a typical shard will be in a production environment:

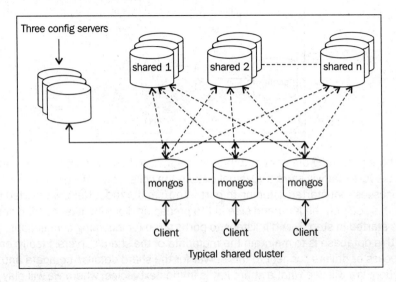

The number of shards will not be two but much more. Also, each shard will be a replica set to ensure high availability. There will be three config servers to ensure the availability of the config servers too. Similarly, there will be any number of `mongos` processes created for a shard that listens for client connections. In some cases, it might even be started on a client application's server.

There's more...

What good is a shard unless we put it to action and see what happens from the shell on inserting and querying the data? In the next recipe, we will make use of the shard setup, add some data, and see it in action.

Connecting to a shard from the Mongo shell and performing operations

In this recipe, we will be connecting to a shard from a command prompt; we will also see how to shard a collection and observe the data splitting in action on some test data.

Getting ready

Obviously, we need a sharded mongo server setup that is up and running. See the previous recipe for more details on how to set up a simple shard. The mongos process, as in the previous recipe, should be listening to port number 27017. We have got some names in a JavaScript file called names.js. This file needs to be downloaded from this book's site and kept on the local filesystem. The file contains a variable called names, and the value is an array with some JSON documents as the values, each one representing a person. The contents look as follows:

```
names = [
  {name:'James Smith', age:30},
  {name:'Robert Johnson', age:22},
  …
]
```

How to do it...

Let's take a look at the steps in detail:

1. Start the Mongo shell and connect to the default port on the localhost as follows (this will ensure that the names will be available in the current shell):

    mongo --shell names.js

    MongoDB shell version: 2.4.6

    connecting to: test

    mongos>

2. Switch to the database that will be used to test sharding as follows (we call it shardDB):

    mongos> use shardDB

3. Enable sharding at the database level as follows:

    mongos> sh.enableSharding("shardDB")

4. Shard a collection called person as follows:

    mongos>sh.shardCollection("shardDB.person", {name: "hashed"}, false)

Installing and Starting the MongoDB Server

5. Add test data to the sharded collection as follows:

    ```
    mongos> for(i = 1; i <= 300000 ; i++) {
    ... person = names[Math.round(Math.random() * 100) % 20]
    ... doc = {_id:i, name:person.name, age:person.age}
    ... db.person.insert(doc)
    }
    ```

6. Execute the following command to get a query plan and the number of documents on each shard:

    ```
    mongos> db.person.find().explain()
    ```

How it works...

This recipe demands some explanation. We have downloaded a JavaScript file that defines an array of 20 people. Each element of the array is a JSON object with a `name` and `age` attribute. We started the shell that connects to the `mongos` process loaded with this JavaScript. We then switched to `shardDB`, which we will use for the purpose of sharding.

For a collection to be sharded, the database in which it will be created needs to be enabled for sharding first. We do this using `sh.enableSharding()`.

The next step is to enable the collection to be sharded. By default, all the data will be kept on one shard and will not be split across different shards. Think about how Mongo will be able to meaningfully split the data. The whole intention is to split it meaningfully and as evenly as possible so that whenever we query based on a shard key, Mongo will easily be able to determine which shard(s) to query. If a query doesn't contain a shard key, the execution of the query will happen on all the shards, and the data will then be collated by the `mongos` process before returning it to the client. Thus, choosing the right shard key is very crucial.

Let's now see how to shard the collection. We will do this by invoking `sh.shardCollection("shardDB.person", {name: "hashed"}, false)`. There are three parameters here.

- The first parameter specifies a fully qualified name of the collection in the `<db name>.<collection name>` format. This is the first parameter of the `shardCollection` method.
- The second parameter specifies the field name to shard upon in the collection. This is the field that will be used to split the documents on the shards. One of the requirements of a good shard key is that it should have high cardinality (the number of possible values should be high). In our test data, the name value has a very low cardinality and thus, is not a good choice as a shard key. We thus hash this key when using it as a shard key. We do so by mentioning the key as `{name: "hashed"}`.

▶ The last parameter specifies whether the value used as a shard key is unique or not. The name field is definitely not unique; thus, it will be false. If the field was, say, the person's **social security number**, it could have been set as true. Also, SSN is a good choice for a shard key due to its high cardinality. Remember though, for the query to be efficient, the shard key has to be present in it.

The last step is to see the execution plan to find all the data. The intent of this operation is to see how the data is being split across two shards. With 3,00,000 documents, we expect something around 1,50,000 documents on each shard. From the explain plan's output, the shard attribute has an array with a document value for each shard in the cluster. In our case. we have two; thus. we have two shards that give the query plan for each shard. In each of them, the value of n is something to look at. It should give us the number of documents that reside on each shard. The following code snippet is the relevant JSON document we see from the console. The number of documents on shards one and two is 164938 and 135062, respectively:

```
"shards" : {
  "localhost:27000" : [
    {
      "cursor" : "BasicCursor",
      "isMultiKey" : false,
      "n" : 164938,
      "nscannedObjects" : 164938,
      "nscanned" : 164938,
      "nscannedObjectsAllPlans" : 164938,
      "nscannedAllPlans" : 164938,
      "scanAndOrder" : false,
      "indexOnly" : false,
      "nYields" : 1,
      "nChunkSkips" : 0,
      "millis" : 974,
      "indexBounds" : {

      },
      "server" : "Amol-PC:27000"
    }
  ],
  "localhost:27001" : [
    {
      "cursor" : "BasicCursor",
      "isMultiKey" : false,
      "n" : 135062,
      "nscannedObjects" : 135062,
      "nscanned" : 135062,
      "nscannedObjectsAllPlans" : 135062,
      "nscannedAllPlans" : 135062,
      "scanAndOrder" : false,
```

Installing and Starting the MongoDB Server

```
            "indexOnly" : false,
            "nYields" : 0,
            "nChunkSkips" : 0,
            "millis" : 863,
            "indexBounds" : {

            },
            "server" : "Amol-PC:27001"
        }
    ]
}
```

There are a couple of additional things that I recommend you all to do.

Connect to the individual shard from the Mongo shell and execute queries on the `person` collection. See that the counts in these collections are similar to what we see in the preceding plan. Also, one can find out that no document exists on both the shards at the same time.

We discussed in brief how cardinality affects the way the data is split across shards. Let's do a simple exercise. We will first drop the `person` collection and execute the `shardCollection` operation again but, this time, with the `{name: 1}` shard key instead of `{name: "hashed"}`. This ensures that the shard key is not hashed and stored as is. Now, load the data using the JavaScript function we used earlier in step 5 and then execute `explain` on the collection once the data is loaded. Observe how the data is now split (or not) across the shards.

There's more...

A lot of questions might now come up, such as what are the best practices, what are some tips and tricks, how is the sharding thing pulled off by MongoDB behind the scenes in a way transparent to the end user, and so on.

This recipe only explained the basics. All these questions will be answered in *Chapter 4, Administration*.

2
Command-line Operations and Indexes

In this chapter, we will cover the following recipes:

- Creating test data
- Performing simple querying, projections, and pagination from the Mongo shell
- Updating and deleting data from the shell
- Creating an index and viewing plans of queries
- Background and foreground index creation from the shell
- Creating unique indexes on collection and deleting the existing duplicate data automatically
- Creating and understanding sparse indexes
- Expiring documents after a fixed interval using the TTL index
- Expiring documents at a given time using the TTL index

Creating test data

This recipe is about creating test data for some of the recipes in this chapter and also for the later chapters in this book. We will demonstrate how to load a CSV file into a Mongo database using the import utility. This is a basic recipe; if readers are aware of the data-import process, they might just download the CSV file (`pincodes.csv`) from the book's site, load it in the collection by themselves, and skip the rest of the recipe. We will use the default database test, and the collection will be named `postalCodes`.

Command-line Operations and Indexes

Getting ready

The data used here is for postal codes in India. Download the `pincodes.csv` file from the book's website. The file is a CSV file with 39,732 records; it should create 39,732 documents upon successful import. We need to have the Mongo server up and running. Refer to the *Single node installation of MongoDB* recipe in *Chapter 1, Installing and Starting the MongoDB Server*, for instructions on how to start the server. The server should begin listening for connections on the default port `27017`.

How to do it...

1. Execute the following command from the shell with the file to be imported in the current directory:

   ```
   $ mongoimport --type csv -d test -c postalCodes --headerline
   --drop pincodes.csv
   ```

2. Start the Mongo shell by typing in `mongo` in the command prompt.
3. In the shell, execute the following command:

   ```
   > db.postalCodes.count()
   ```

How it works...

Assuming that the server is up and running, the CSV file is downloaded and kept in a local directory where we can execute the import utility with the file in the current directory. Let's look at the options given to the Mongo import utility and their meanings:

Command-line option	Decription
`--type`	This specifies that the type of input file is CSV. It defaults to JSON, the other possible value being TSV.
`-d`	This is the target database into which the data will be loaded.
`-c`	This is the collection in the preceding database into which the data will be loaded.
`--headerline`	This is relevant only in the case of TSV or CSV files. It indicates that the first line of the file is the header. The same name will be used as the name of the field in the document.
`--drop`	This indicates that we need to drop the collection before the data gets loaded in it.

After all the options are given. the final value in the command prompt is the name of the file, `pincodes.csv`.

Chapter 2

If the import goes through successfully, you will see something like the following output on the console:

```
connected to: 127.0.0.1
Mon Dec  9 23:29:13.004     Progress: 1593394/2286080        69%
Mon Dec  9 23:29:13.014     28000 9333/second
Mon Dec  9 23:29:14.116     check 9 39733
Mon Dec  9 23:29:14.116     imported 39732 objects
```

Finally, we will start the Mongo shell and find the count of the documents in the collection. It should indeed be 39,732, as seen in the preceding import log.

The postal code data is taken from https://github.com/kishorek/India-Codes/. This data is not taken from an official source and might not be accurate as it is being compiled manually for free public use. Thanks to Kishore for compiling the data and sharing it.

See also

- The *Performing simple querying, projections, and pagination from the Mongo shell* recipe to know some basic queries on the data imported

Performing simple querying, projections, and pagination from the Mongo shell

In this recipe, we will get our hands dirty with a bit of querying to select documents from the test data we set up in the previous recipe. There is nothing extravagant in this recipe, and someone well versed with query language basics can skip this recipe comfortably. Others who aren't too comfortable with basic querying or those who want to get a small refresher can continue to read the next section of the recipe. Additionally, this recipe is intended to give a feel of the test data setup from the previous recipe.

Getting ready

To execute simple queries, we need to have a server up and running. A simple single node is what we will need. Refer to the *Single node installation of MongoDB* recipe in *Chapter 1, Installing and Starting the MongoDB Server*, to learn how to start the server. The data on which we will be operating needs to be imported into the database. The steps to import the data are given in the previous recipe. You also need to start the Mongo shell and connect to the server that runs on the localhost. Once we have these prerequisites, we are good to go.

Command-line Operations and Indexes

How to do it...

1. Let's first find a count of documents in the collection:

   ```
   > db.postalCodes.count()
   ```

2. Let's find just one document from the `postalCodes` collection:

   ```
   > db.postalCodes.findOne()
   ```

3. Now, we need to find multiple documents in the collection:

   ```
   > db.postalCodes.find().pretty()
   ```

4. The preceding query retrieved all the keys of the first 20 documents and displayed them on the shell. Let's do a couple of things now; we will just display the `city`, `state`, and `pincode` fields. Additionally, we want to display the documents numbered 91 to 100 in the collection. To do this, execute the following command:

   ```
   > db.postalCodes.find({}, {_id:0, city:1, state:1, pincode:1}).skip(90).limit(10)
   ```

5. Let's move a step ahead and write a slightly complex query where we will find the top 10 cities in the state of Gujarat sorted by the name of the city. Like the last query, we will just select the `city`, `state`, and `pincode` fields:

   ```
   > db.postalCodes.find({state:'Gujarat'},{_id:0, city:1, state:1, pincode:1}).sort({city:1}).limit(10)
   ```

How it works...

The recipe is pretty simple and allows us to get a feel of the test data we set up in the previous recipe. Nevertheless, as with the other recipes, I owe you all some explanation about what we did here.

We first found the count of the documents in the collection using `db.postalCodes.count()`, and it should give us 39,732 documents. This should be in sync with the logs we saw while importing the data into the `postalCodes` collection. Next, we queried for one document from the collection using `findOne`. This method returned the first document in the result set of the query. In the absence of a query or sort order, as in this case, it will be the first document in the collection sorted by its natural order.

Next, we used `find` rather than `findOne`. The difference is that the `find` operation returns an iterator for the result set that we can use to traverse through the results of the `find` operation, whereas `findOne` returns a document. Adding a `pretty` method call to the `find` operation will print the result in a pretty or formatted way.

> Note that the `pretty` method makes sense and works only with the `find` method and not with `findOne`. This is because the return value of `findOne` is a document, and there is no `pretty` operation on the returned document.

We will now execute the following query on the Mongo shell:

```
> db.postalCodes.find({}, {_id:0, city:1, state:1, pincode:1}).skip(90).limit(10)
```

Here, we will pass two parameters to the `find` method:

- The first one is `{ }`, which is the query to select the documents; in this case, we will ask Mongo to select all the documents.
- The second parameter is the set of fields that we want in the result documents. Remember that the `_id` field is present by default, unless we explicitly say `_id:0`. For all the other fields, we need to say `<field_name>:1` or `<field_name>:true`. The `find` method with projections is the same as saying select field 1 and field 2 from the table in the relational world, and not specifying the fields to be selected in the `find` method is like saying select * from the table in the relational world.

Moving on, we just need to look at what `skip` and `limit` do. The `skip` function skips the given number of documents from the result set, all the way up to the end document in the result set. The `limit` function then limits the result to the given number of documents.

Let's see what all this means with an example. By executing `.skip(90).limit(10)`, we say that we want to skip the first 90 documents from the result set and start returning from the ninety-first document. The limit, however, says that we will return only 10 documents from the ninety-first document.

Now, there are some border conditions that we need to know here. What if `skip` is being provided with a value more than the total number of documents in the collection? Well, in this case, no documents will be returned. Also, if the number provided to the `limit` function is more than the actual number of documents that remain in the collection, the number of documents returned will be the same as the remaining documents in the collection, and no exception will be thrown in either case.

Updating and deleting data from the shell

This, again, will be a simple recipe that will look at executing deletes and updates on a test collection. We won't deal with the same test data we imported, as we don't want to update/delete any of this; instead, we will work on a test collection created only for this recipe.

Getting ready

For this recipe, we will create a collection called `updAndDelTest`. We will need the server to be up and running. Refer to the *Single node installation of MongoDB* recipe in *Chapter 1, Installing and Starting the MongoDB Server*, to learn how to start the server. Also, start the shell with the `UpdAndDelTest.js` script loaded. This script will be available on the book's website for download. To know how to start the shell with a reloaded script, refer to the *Connecting to a single node from the Mongo shell with a preloaded JavaScript* recipe in *Chapter 1, Installing and Starting the MongoDB Server*.

How to do it...

1. With the shell started and the script loaded, execute the following command in the shell:

    ```
    > prepareTestData()
    ```

    If all goes well, you should see 20 documents inserted in `updAndDelTest` and printed on the console.

2. To get a feel of the collection, let's query it as follows:

    ```
    > db.updAndDelTest.find({}, {_id:0})
    ```

3. We will see that for each value of x as 1 and 2, we have y incrementing from 1 to 10.
4. We will first update some documents and observe the results. Execute the following update command:

    ```
    > db.updAndDelTest.update({x:1}, {$set:{y:0}})
    ```

5. Execute the following `find` command and observe the results; we will get 10 documents (for each of them, note the value of y):

    ```
    > db.updAndDelTest.find({x:1}, {_id:0})
    ```

6. We will now execute the following `update` command:

    ```
    > db.updAndDelTest.update({x:1}, {$set:{y:0}}, false, true)
    ```

7. Executing the query given in step 5 again to view the updated documents will show the same documents we saw earlier. Take a note of the values of y again and compare them to the results we saw when we executed this query the last time around before executing the update command given in step 6.

8. We will now see how the delete operation works. We will again choose the documents where x is 1 for the deletion test. Let's delete all the documents where x is 1 from the collection:

    > db.updAndDelTest.remove({x:1})

9. Execute the following find command and observe the results. We will not get any results. It seems that the remove operation has removed all the documents with x as 1:

    > db.updAndDelTest.find({x:1}, {_id:0})

How it works...

First, we set up the data that we would be using for updates and deletion. We have already seen the data and know what it is. An interesting thing to observe is that, when we execute an update such as db.updAndDelTest.update({x:1}, {$set:{y:0}}), it only updates the first document that matches the query provided as the first parameter. This is something we will observe when we query the collection after this update. The update function has the db.<collection name>.update(query, update object, isUpsert, isMulti) format.

We will see what Upsert is in *Atomic find and modify operations* recipe in *Chapter 5, Advanced Operations*. The fourth parameter (isMulti) is by default false, and this means that multiple documents will not be updated by the update call. So, only the first matching document will be updated by default. However, when we execute db.updAndDelTest.update({x:1}, {$set:{y:0}}, false, true) with the fourth parameter set to true, all the documents in the collection that match the given query get updated. This is something we can verify after querying the collection.

Removals, on other hand, behave differently. By default, the remove operation deletes all the documents that match the provided query. However, if we wish to delete only one document, we will explicitly pass the second parameter as true.

> The default behavior of update and remove is different. An update call by default updates only the first matching document, whereas remove deletes all the documents that match the query.

Creating an index and viewing plans of queries

In this recipe, we will look at querying data, analyzing its performance by explaining the query plan, and then optimizing it by creating indexes.

Getting ready

For the creation of indexes, we need to have a server up and running. A simple single node is what we will need. Refer to the *Single node installation of MongoDB* recipe in *Chapter 1, Installing and Starting the MongoDB Server*, to learn how to start the server. The data with which we will be operating needs to be imported into the database. The steps to import the data are given in the *Creating test data* recipe. Once we have this prerequisite, we are good to go.

How to do it...

We will trying to write a query that will find all the zip codes in a given state. To do this, perform the following steps:

1. Execute the following query to view the plan of a query:

   ```
   > db.postalCodes.find({state:'Maharashtra'}).explain()
   ```

   Take a note of the `cursor`, `n`, `nscannedObjects`, and `millis` fields in the result of the `explain` plan operation

2. Let's execute the same query again; this time, however, we will limit the results to only 100 results:

   ```
   > db.postalCodes.find({state:'Maharashtra'}).limit(100).explain()
   ```

   Again, take a note of the cursor, `n`, `nscannedObjects`, and `millis` fields in the result

3. We will now create an index on the `state` and `pincode` fields as follows:

   ```
   > db.postalCodes.ensureIndex({state:1, pincode:1})
   ```

4. Execute the following query:

   ```
   > db.postalCodes.find({state:'Maharashtra'}).explain()
   ```

   Again, take a note of the `cursor, n, nscannedObjects, millis,` and `indexOnly` fields in the result

Chapter 2

5. Since we want only the pin codes, we will modify the query as follows and view its plan:

    ```
    > db.postalCodes.find({state:'Maharashtra'}, {pincode:1, _id:0}).explain()
    ```

    Take a note of the `cursor`, `n`, `nscannedObjects`, `nscanned`, `millis`, and `indexOnly` fields in the result.

How it works...

There is a lot to explain here. We will first discuss what we just did and how to analyze the stats. Next, we will discuss some points to be kept in mind for index creation and some gotchas.

Analyzing the plan

Let's look at the first step and analyze the output we executed:

```
> db.postalCodes.find({state:'Maharashtra'}).explain()
```

The output on my machine is as follows (I am skipping the nonrelevant fields for now):

```
{
  "cursor" : "BasicCursor",
  ...
  "n" : 6446,
  "nscannedObjects" : 39732,
  "nscanned" : 39732,
  ...
  "millis" : 55,
  ...
}
```

The value of the `cursor` field in the result is `BasicCursor`, which means a full collection scan (all the documents are scanned one after another) has happened to search the matching documents in the entire collection. The value of `n` is `6446`, which is the number of results that matched the query. The `nscanned` and `nscannedobjects` fields have values of 39,732, which is the number of documents in the collection that are scanned to retrieve the results. This is the also the total number of documents present in the collection, and all were scanned for the result. Finally, `millis` is the number of milliseconds taken to retrieve the result.

49

Improving the query execution time

So far, the query doesn't look too good in terms of performance, and there is great scope for improvement. To demonstrate how the limit applied to the query affects the query plan, we can find the query plan again without the index but with the `limit` clause:

```
> db.postalCodes.find({state:'Maharashtra'}).limit(100).explain()

{
    "cursor" : "BasicCursor",
    ...
    "n" : 100,
    "nscannedObjects" : 19951,
    "nscanned" : 19951,
    ...
    "millis" : 30,
    ...
}
```

The query plan this time around is interesting. Though we still haven't created an index, we saw an improvement in the time the query took for execution and the number of objects scanned to retrieve the results. This is due to the fact that Mongo does not scan the remaining documents once the number of documents specified in the `limit` function is reached. We can thus conclude that it is recommended that you use the `limit` function to limit your number of results, whereas the maximum number of documents accessed is known upfront. This might give better query performance. The word "might" is important as, in the absence of an index, the collection might still be completely scanned if the number of matches is not met.

Improvement using indexes

Moving on, we will create a compound index on `state` and `pincode`. The order of the index is ascending in this case (as the value is 1) and is not significant unless we plan to execute a multikey sort. This is a deciding factor as to whether the result can be sorted using only the index or whether Mongo needs to sort it in memory later on, before we return the results. As far as the plan of the query is concerned, we can see that there is a significant improvement:

```
{
    "cursor" : "BtreeCursor state_1_pincode_1",
    ...
    "n" : 6446,
    "nscannedObjects" : 6446,
    "nscanned" : 6446,
```

```
    ...
    "indexOnly" : false,
    ...
    "millis" : 16,
    ...
}
```

The `cursor` field now has the `BtreeCursor state_1_pincode_1` value, which shows that the index is indeed used now. As expected, the number of results stays the same at `6446`. The number of objects scanned in the index and documents scanned in the collection have now reduced to the same number of documents as in the result. This is because we have now used an index that gave us the starting document from which we could scan; then, only the required number of documents was scanned. This is similar to using the book's index to find a word or scanning the entire book to search for the word. The time, `millis`, has come down too, as expected.

Improvement using covered indexes

This leaves us with one field, `indexOnly`, and we will see what this means. To know what this value is, we need to look briefly at how indexes operate.

Indexes store a subset of fields of the original document in the collection. The fields present in the index are the same as those on which the index is created. The fields, however, are kept sorted in the index in an order specified during the creation of the index. Apart from the fields, there is an additional value stored in the index; this acts as a pointer to the original document in the collection. Thus, whenever the user executes a query, if the query contains fields on which an index is present, the index is consulted to get a set of matches. The pointer stored with the index entries that match the query is then used to make another I/O operation to fetch the complete document from the collection; this document is then returned to the user.

The value of `indexOnly`, which is `false`, indicates that the data requested by the user in the query is not entirely present in the index; an additional I/O operation is needed to retrieve the entire document from the collection that follows the pointer from the index. Had the value been present in the index itself, an additional operation to retrieve the document from the collection would not be necessary, and the data from the index will be returned. This is called a covered index, and the value of `indexOnly`, in this case, will be `true`.

Command-line Operations and Indexes

In our case, we just need the pin codes, so why not use projection in our queries to retrieve just what we need? This will also make the index covered as the index entry that just has the state's name and pin code, and the required data, can be served completely without retrieving the original document from the collection. The plan of the query in this case is interesting too. Executing the following query plan:

```
db.postalCodes.find({state:'Maharashtra'}, {pincode:1, _id:0}).explain()
{
    "cursor" : "BtreeCursor state_1_pincode_1",
    ...
    "n" : 6446,
    "nscannedObjects" : 0,
    "nscanned" : 6446,
    ...
    "indexOnly" : true,
    ...
    "millis" : 15,
    ...
}
```

The values of the `nscannedobjects` and `indexOnly` fields are something to be observed. As expected, since the data we requested in the projection in the `find` query is the pin code only, which can be served from the index alone, the value of `indexOnly` is `true`. In this case, we scanned 6,446 entries in the index; thus, the `nscanned` value is `6446`. We, however, didn't reach out to any document in the collection on the disk, as this query was covered by the index alone, and no additional I/O was needed to retrieve the entire document. Hence, the value of `nscannedobjects` is 0.

As this collection in our case is small, we do not see a significant difference in the execution time of the query. This will be more evident on larger collections. Making use of indexes is great and gives good performance. Making use of covered indexes gives even better performance.

Another thing to remember is that, wherever possible, try and use projection to retrieve only the number of fields we need. The `_id` field is retrieved every time by default, unless we plan to set `_id:0` to not retrieve it if it is not part of the index. Executing a covered query is the most efficient way to query a collection.

Some gotchas of index creation

We will now see some pitfalls in index creation and some facts about the array field, which is used in the index.

Some of the operators that do not use the index efficiently are the $where, $nin, and $exists operators. Whenever these operators are used in the query, one should bear in mind a possible performance bottleneck when the data size increases. Similarly, the $in operator must be preferred over the $or operator, as both can be more or less used to achieve the same result. As an exercise, try to find the pin codes in the state of Maharashtra and Gujarat from the postalCodes collection. Write two queries: one using the $or operator and the other using the $in operator. Explain the plan for both these queries.

What happens when an array field is used in the index? Mongo creates an index entry for each element present in the array field of a document. So, if there are 10 elements in an array in a document, there will be 10 index entries, one for each element in the array. However, there is a constraint while creating indexes that contain array fields. When creating indexes using multiple fields, no more than one field can be of the array type. This is done to prevent a possible explosion in the number of indexes on adding even a single element to the array used in the index. If we think about it carefully, for each element in the array, an index entry is created. If multiple fields of type array were allowed to be part of an index, we would have a large number of entries in the index that would be a product of the length of these array fields. For example, a document added with two array fields, each of length 10, will add 100 entries to the index, had it been allowed to create one index using these two array fields.

This should be good enough for now to scratch the surfaces of a plain vanilla index. We will see more options and types in some of the upcoming recipes.

Background and foreground index creation from the shell

In the previous recipe, we looked at how to analyze queries, how to decide what index needs to be created, and how we create indexes. This, by itself, is straightforward and looks reasonably simple. However, for large collections, things start getting worse as the index-creation time is large. There are some caveats that we need to keep in mind. The objective of this recipe is to throw some light on these concepts and avoid pitfalls while creating indexes, especially on large collections.

Command-line Operations and Indexes

Getting ready

For the creation of indexes, we need to have a server up and running. A simple single node is what we will need. Refer to the *Single node installation of MongoDB* recipe in *Chapter 1, Installing and Starting the MongoDB Server,* for how to start the server.

Start connecting two shells to the server by just typing in `mongo` from the operating system shell. Both of them will, by default, connect to the `test` database.

Our test data for zipped codes is pretty small to demonstrate the problem faced during index creation on large collections. We need to have more data; thus, we will start by creating some to simulate the problems during index creation. The data has no practical meaning but is good enough to test the concepts. Copy the following piece of code in one of the started shells and execute it (it is a pretty easy snippet to type out too):

```
for(i = 0; i < 5000000 ; i++) {
  doc = {}
  doc._id = i
  doc.value = 'Some text with no meaning and number ' + i + ' in between'
  db.indexTest.insert(doc)
}
```

A document in this collection will be as follows:

```
{ _id:0, value:"Some text with no meaning and number 0 in between" }
```

Execution will take a quite a lot of time, so we need to be patient. Once the execution is over, we are all set for the action.

> If you are keen to know what the current number of documents loaded in the collection is, evaluate the following command from the second shell periodically:
>
> `> db.indexTest.count()`

How to do it...

1. Create an index on the `value` field of the document:

   `> db.indexTest.ensureIndex({value:1})`

2. While the index creation is in progress, which will take quite some time, switch over to the second console and execute the following command:

   `> db.indexTest.findOne()`

   Both the index creation shell and the one where we executed findOne will be blocked, and the prompt will not be shown on both of them until the index creation is complete.

3. Now, this was foreground index creation by default. We want to see the behavior in background index creation. Drop the created index:

   ```
   > db.indexTest.dropIndex({value:1})
   ```

4. Create the index again but, this time, in the background:

   ```
   > db.indexTest.ensureIndex({value:1}, {background:true})
   ```

5. In the second Mongo shell, execute the `findOne` query this time around:

   ```
   > db.indexTest.findOne()
   ```

   This should return one document this time around, unlike the first instance where the operation was blocked until index creation completed in the foreground

6. In the second shell, also repeatedly execute the following `explain` operation with an interval of about 4 to 5 seconds between each `explain` plan invocation until the index-creation process is complete:

   ```
   > db.indexTest.find({value:"Some text with no meaning and number 0 in between"}).explain()
   ```

How it works...

Let's now analyze what we just did. We created about 5 million documents with no practical importance, but we are just looking to get some data that will take a significant amount of time for index building.

Indexes can be built in two ways, in the foreground and background. In either case, the shell doesn't show the prompt until the `ensureIndex` operation is completed and it doesn't show the blocks till the index is created. You might then be wondering what difference it makes to create an index in the background or foreground.

That is exactly where the second shell we started came into the picture. This is where we demonstrated the difference between a background and foreground index-creation process. We first created the index in the foreground, which is the default behavior. This index building didn't allow us to query the collection (from the second shell) until the index was constructed. The `findOne` operation is blocked until the entire index is built (from the first shell) before returning the result. On the other hand, the index that was built in the background didn't block the `findOne` operation. If you want to try inserting new documents into the collection while the index build is on, this too should work well. Feel free to drop the index and recreate it in the background while simultaneously inserting a document in the `indexTest` collection; you will notice that it works smoothly.

Well, what is the difference between the two approaches and why not always build the index in the background? Apart from an extra parameter, `{background:true}`, which can also be `{background:1}`, passed as a second parameter to the `ensureIndex` call, there are few differences. The index-creation process in the background will be slightly slower than the index created in the foreground. Furthermore, internally, though it is not relevant to the end user, the index created in the foreground will be more compact than the one created in the background.

Other than that, there will be no significant difference. In fact, if a system is running and an index needs to be created while it is serving the end users (not recommended, but there can be a situation that demands index creation on a live system), then creating the index in the background is the only way we can do it. There are other strategies for performing such administrative activities that we will see in some recipes in *Chapter 4, Administration*.

To make things worse for foreground index creation, the lock acquired by Mongo during index creation is not at the collection level but at the database level. To explain what this means, we will have to drop the index on the `indexTest` collection and perform a small exercise as follows:

1. Start by creating the index in the foreground from the shell by executing the following command:

    ```
    > db.indexTest.ensureIndex({value:1})
    ```

2. Now, insert a document in the `person` collection, which might or might not exist at this point in the `test` database:

    ```
    > db.person.insert({name:'Amol'})
    ```

We will see that this `insert` operation on the `person` collection will create a block, while the index creation on the `indexTest` collection is in process. If, however, this `insert` operation is done on a collection in a different database during index build (you can try this out too), it will execute normally without blocking. This clearly shows that the lock is acquired at the database level and not at the collection level or global level.

> Prior to version 2.2 of Mongo, locks were at the global level, which is at the `mongod` process level and not at the database level as we saw earlier. You need to remember this fact when dealing with the distribution of Mongo that is older than version 2.2.

Creating unique indexes on collection and deleting the existing duplicate data automatically

In this recipe, we will look at creating unique indexes on a collection. Unique indexes, from the name itself, tell us that the value with which the index is created has to be unique. What if the collection already has data and we want to create a unique index on a field whose value is not unique in the existing data?

Obviously, we cannot create the index, and it will fail. There is, however, a way to drop the duplicates and create the index. Curious how this can be achieved? Yes? Keep reading this recipe.

Getting ready

For this recipe, we will create a collection called `userDetails`. We will need the server to be up and running. Refer to the *Single node installation of MongoDB* recipe in *Chapter 1, Installing and Starting the MongoDB Server*, to learn how to start the server. Also, start the shell with the `UniqueIndexData.js` script loaded. This script will be available on the book's website for download. To find out how to start the shell with a script reloaded, refer to the *Connecting to a single node from the Mongo shell with a preloaded JavaScript* recipe in *Chapter 1, Installing and Starting the MongoDB Server*.

How to do it...

1. Load the required data in the collection using the `loadUserDetailsData` method.
2. Execute the following command on the Mongo shell:

    ```
    > loadUserDetailsData()
    ```

3. See the count of the documents in the collection using the following query (it should be 100):

    ```
    > db.userDetails.count()
    ```

4. Now, try to create a unique index on the `login` field on the `userDetails` collection:

    ```
    > db.userDetails.ensureIndex({login:1}, {unique:true})
    ```

Command-line Operations and Indexes

5. This will not be successful and something like the following error will be seen on the console:

   ```
   {
       "err" : "E11000 duplicate key error index: test.userDetails.$login_1  dup key: { : \"bander\" }",
       "code" : 11000,
       "n" : 0,
       "connectionId" : 6,
       "ok" : 1
   }
   ```

6. Next, we will try to create an index on this collection by eliminating the duplicates:

   ```
   > db.userDetails.ensureIndex({login:1}, {unique:true, dropDups:true})
   ```

7. This will throw no errors and find the count in the collection again (take a note of the count and compare it with the count seen earlier, prior to index creation):

   ```
   > db.userDetails.count()
   ```

8. Check whether the index is being used by viewing the plan of the query:

   ```
   > db.userDetails.find({login:'mtaylo'}).explain()
   ```

How it works...

We initially loaded our collection with 100 documents using the `loadUserDetailsData` function from the `UniqueIndexData.js` file. We looped 100 times and loaded the same data over and over again. Thus, we got duplicate documents.

We will then try to create a unique index on the `login` field in the `userDetails` collection as follows:

```
> db.userDetails.ensureIndex({login:1}, {unique:true})
```

This creation fails and indicates the duplicate key it first encountered on index creation. It is `bander` in this case. Can you guess why an error was first encountered for this user ID? This is not even the first ID we saw in the loaded data.

 When specifying 1 in index creation, we mean to convey that the order of the values is ascending. Try creating a unique index using `{login:-1}` and see if the user ID for which the error is encountered is different.

In such a scenario, we are left with two options:

- Manually pick the data to be deleted/fixed and ensure that the field on which the index is to be created has unique data across collection. This can either be done manually or programmatically, but it is outside the scope of Mongo and done by the end user on a case-to-case basis.

- Alternatively, if we don't care much about the data as it is genuinely duplicated and we need to retain just one copy of it, Mongo provides a brilliant way to handle this. Apart from the regular {unique:true} option used to create a unique index, we will provide an additional dropDups:true option (or dropDups:1 if you wish) that will blindly delete all the duplicate data it encounters during index creation. Note that there is no guarantee of which document will be retained and which one will be deleted, but just one will be retained. In this case, there are 20 unique login IDs. On unique index creation, if the value of the login ID is not already present in the index, it will be added. Subsequently, when the login ID encountered is already present in the index, the corresponding document is deleted from the collection; this explains why we were left with just 20 documents in the userDetails collection.

Creating and understanding sparse indexes

Schema-free design is one of the fundamental features of Mongo. This allows documents in a collection to have disparate fields, with some fields present in some documents and absent in the others. In other words, these fields might be sparse; this might have already given you a clue to what sparse indexes are. In this recipe, we will create some random test data and see how sparse indexes behave against a normal index. We will see the advantage of using a sparse index and one major pitfall in its usage.

Getting ready

For this recipe, we need to create a collection called sparseTest. We will require a server to be up and running. Refer to the *Single node installation of MongoDB* recipe in *Chapter 1, Installing and Starting the MongoDB Server*, to learn how to start the server. Also, start the shell with the SparseIndexData.js script loaded. This script will be available on the book's website for download. To know how to start the shell with a script reloaded, refer to the *Connecting to a single node from the Mongo shell with a preloaded JavaScript* recipe in *Chapter 1, Installing and Starting the MongoDB Server*.

How to do it...

1. Load the data in the collection by invoking the following command (this should import 100 documents in the `sparseTest` collection):

   ```
   > createSparseIndexData()
   ```

2. Now, take a look at the data by executing the following query, taking note of the `y` field in the top few results:

   ```
   > db.sparseTest.find({}, {_id:0})
   ```

3. We can see that the `y` field is either absent, or it is unique if it is present. Let's then execute the following query:

   ```
   > db.sparseTest.find({y:{$ne:2}}, {_id:0}).limit(15)
   ```

4. Take a note of the result; it contains both the documents that match the condition as well as the fields that do not contain the given `y` field.

5. Since the value of `y` seems unique, let's create a new unique index on the `y` field:

   ```
   > db.sparseTest.ensureIndex({y:1}, {unique:1})
   ```

6. This throws an error; it complains that the value is not unique and that the offending value is the null value.

7. We will fix this by making this index sparse as follows:

   ```
   > db.sparseTest.ensureIndex({y:1}, {unique:1, sparse:1})
   ```

8. This should fix our problem. To confirm that the index got created, execute the following command on the shell:

   ```
   > db.sparseTest.getIndexes()
   ```

9. This should show two indexes: the default one on `_id` and the one we just created in the preceding step.

10. Now, execute the query we executed in step 3 again, and see the result.

11. Look at the result and compare it with what we saw before the index was created. Re-execute the query but with the following hint, forcing a full table scan:

    ```
    > db.sparseTest.find({y:{$ne:2}},{_id:0}).limit(15).hint({$natural:1})
    ```

12. Observe the result again.

How it works...

Those were a lot of steps and work that we did. We will now dig deeper and explain the internals and the reasoning for the weird behavior we saw while querying the collection that used sparse indexes.

The test data that we created using the JavaScript method just created documents with an x key whose value is a number starting from 1 and can go all the way up to 100. The value of y is set only when x is a multiple of 3; its value too is a running number starting from 1 and should go up to a maximum of 33 when x is 99.

We will then execute the following query and see the following result as expected:

```
> db.sparseTest.find({y:{$ne:2}}, {_id:0}).limit(15)
{ "x" : 1 }
{ "x" : 2 }
{ "x" : 3, "y" : 1 }
{ "x" : 4 }
{ "x" : 5 }
{ "x" : 7 }
{ "x" : 8 }
{ "x" : 9, "y" : 3 }
{ "x" : 10 }
{ "x" : 11 }
{ "x" : 12, "y" : 4 }
{ "x" : 13 }
{ "x" : 14 }
{ "x" : 15, "y" : 5 }
{ "x" : 16 }
```

The value where y is 2 is missing in the result, and this is what we intended. Note that the documents where y isn't present are still seen in the result. We will now plan to create an index on the field y. As the field is either not present or has a value that is unique, it seems natural that a unique index should work.

Internally, indexes, by default, add an entry in the index even if the field is absent in the original document in the collection. The value that goes in the index will however be null. This means that there will be the same number of entries in the index as the number of documents in the collection. For a unique index, the value (including null values) should be unique across the collection; this explains why we got an exception during index creation where the field is sparse (not present in all documents).

A solution for this problem is to make the index sparse, and all we did was add `sparse:1` to the options along with `unique:1`. This does not put an entry in the index if the field doesn't exist in the document. Thus, the index will now contain fewer entries. It will only contain those entries where the field is present in the document. This not only makes the index smaller, making it easy to fit in memory, but also solves our problem of adding a unique constraint. The last thing we want is to have an index of a collection with millions of documents with millions of entries where only a few hundred have some values defined.

Though we saw that creating a sparse index made the index efficient, it introduced a new problem where some query results were not consistent. When we executed the same query earlier, it yielded different results. Execute the following command:

```
> db.sparseTest.find({y:{$ne:2}}, {_id:0}).limit(15)
{ "x" : 3,  "y" : 1 }
{ "x" : 9,  "y" : 3 }
{ "x" : 12, "y" : 4 }
{ "x" : 15, "y" : 5 }
{ "x" : 18, "y" : 6 }
{ "x" : 21, "y" : 7 }
{ "x" : 24, "y" : 8 }
{ "x" : 27, "y" : 9 }
{ "x" : 30, "y" : 10 }
{ "x" : 33, "y" : 11 }
{ "x" : 36, "y" : 12 }
{ "x" : 39, "y" : 13 }
{ "x" : 42, "y" : 14 }
{ "x" : 45, "y" : 15 }
{ "x" : 48, "y" : 16 }
```

Why did this happen? The answer lies in the query plan for this query. Execute the following command to view the plan of this query:

```
> db.sparseTest.find({y:{$ne:2}}, {_id:0}).limit(15).explain()
```

The plan shows that it used the index to fetch the matching results. As this is a sparse index, all the documents that didn't have the field `y` are not present in it; this didn't show up in the result though it should have. This is the pitfall we need to be careful of when querying a collection with a sparse index and the query happens to use the index. It will yield unexpected results. One solution is to force a full table scan, where we provide the query analyzer with a hint, using the `hint` function.

Chapter 2

Hints are used to force query analyzers to use a user-specified index. Though this is usually not recommended as you really need to know what you are doing, this is one of the scenarios where this is really needed. So, how do we force a full table scan? All we need to do is provide `{$natural:1}` in the `hint` function. The natural ordering of a collection is the order in which it is stored on the disk for a particular collection. This hint forces a full table scan; now, we will get the results as we did earlier. The query performance will, however, degrade for large collections, as it is now using a full table scan.

If the field is present in a lot of documents (there is no formal cut off for what is a lot; it can be 50 percent for some or 75 percent for others) and not really sparse, making the index sparse doesn't make much sense, apart from when we want to make it unique.

> Remember that the null value of a field and the one not present in the document are different. If two documents have a null value for the same field, unique index creation will fail and creating it as sparse index will not help either.

Expiring documents after a fixed interval using the TTL index

One of the nice and interesting features in Mongo is automatically expiring data in the collection after a predetermined amount of time. This is a very useful tool when we desire to purge some data older than a particular timeframe. For a relational database, it is not common for folks to set up a batch job that runs every night to perform this operation.

With the **Time To Live** (**TTL**) feature of Mongo, we need not worry about this as the database takes care of it out-of-the-box. Let's see how we can achieve this.

Getting ready

Let's create some data in Mongo that we want to play with using the TTL indexes. We will create a collection called `ttlTest` for this purpose. We will require a server to be up and running. Refer to the *Single node installation of MongoDB* recipe in *Chapter 1, Installing and Starting the MongoDB Server*, to learn how to start the server. Also, start the shell with the `TTLData.js` script loaded. This script will be available on the book's website for download. To know how to start the shell with a script reloaded, refer to the *Connecting to a single node from the Mongo shell with a preloaded JavaScript* recipe in *Chapter 1, Installing and Starting the MongoDB Server*.

How to do it...

1. Assuming that the server is started and the script provided is loaded on the shell, invoke the following method from the Mongo shell:

   ```
   > addTTLTestData()
   ```

2. Create a TTL index on the `createDate` field:

   ```
   > db.ttlTest.ensureIndex({createDate:1}, {expireAfterSeconds:300})
   ```

3. Now, query the collection:

   ```
   > db.ttlTest.find()
   ```

4. This should give three documents. Repeat the process and execute the `find` query in approximately 30 to 40 seconds repeatedly, to see the three documents getting deleted until the entire collection has zero documents left in it.

How it works...

Let's start by opening the `TTLData.js` file and see what is going on in it. The code is pretty simple; it just got the current date using `new Date()`. It then created three documents with `createDate` that were some 4, 3, and 2 minutes behind the current time for the three documents. So, on the execution of the `addTTLTestData()` method in this script, we will have three documents in the `ttlTest` collection, each having a difference of 1 minute in their creation time.

The next step is the core of the TTL feature: the creation of the TTL index. It is similar to the creation of any other index using the `ensureIndex` method, except that it also accepts a second parameter, a JSON object. Let's see what these two parameters are:

- The first parameter is `{createDate:1}`; this will tell Mongo to create an index on the `createDate` field, and the order of the index is ascending as the value is 1 (-1 would have been descending)
- The second parameter, `{expireAfterSeconds:300}`, is what makes this index a TTL index; it tells Mongo to automatically expire the documents after 300 seconds (5 minutes)

OK, but 5 minutes since when? Since the time they were inserted in the collection or is it some other timestamp? In this case it considers the `createTime` field as the base, as this was the field on which we created the index.

This now raises a question: if a field is being used as the base for the computation of time, there has to be some restriction on its type. It just doesn't make sense to create a TTL index, as we created earlier, on a `char` field that holds, say, the name of a person.

Chapter 2

As we guessed, the type of the field can be a BSON type date or an array of dates. What will happen in the case where an array has multiple dates? What will be considered in this case?

It turns out that Mongo uses the minimum of dates available in the array. Try out this scenario as an exercise.

Put two dates separated by about 5 minutes from each other in a document against the `updateField` field name and then create a TTL index on this field, as you did earlier, to expire the document after 10 minutes (600 seconds). Query the collection and see when the document gets deleted from the collection. It should get deleted after roughly 10 minutes have elapsed since the minimum time value present in the `updateField` array.

Apart from the constraint for the type of field, there are a few more constraints.

- If a field already has an index on it, you cannot create a TTL index on it. As the `_id` field of the collection already has an index by default, it effectively means you cannot create a TTL index on the `_id` field.
- A TTL index cannot be a compound index that involves multiple fields.
- If a field doesn't exist, it will never expire (this is pretty logical, I guess).
- A TTL index cannot be created on capped collections. In case you are not aware of capped collections, they are special collections in Mongo with a size limit on them with the **first in first out** (**FIFO**) insertion order; they delete old documents to make place for new documents, if needed.

 TTL indexes are supported only on Mongo Version 2.2 and above. Also note that the document will not be deleted at exactly the given time in the field. The cycle will be of the granularity of 1 minute; it will delete all the documents eligible for deletion since the last time the cycle was run.

There's more...

A use case might not demand the deletion of all the documents after a fixed interval has elapsed. What if we want to customize the point until which a document stays in the collection? This too can be achieved, and will be demonstrated in the next recipe.

Expiring documents at a given time using the TTL index

In the previous recipe, we saw how documents can be expired after a fixed time period. However, there can be some cases where we might want to have documents that expire at different times. This is not what we saw in the previous recipe. In this recipe, we will see how we can specify the time at which a document can be expired (it might be different for different documents).

Getting ready

For this recipe, we will create a collection called `ttlTest2`. We will require a server to be up and running. Refer to the *Single node installation of MongoDB* recipe in *Chapter 1, Installing and Starting the MongoDB Server*, to learn how to start the server. Also, start the shell with the `TTLData.js` script loaded. This script will be available on the book's website for download. To know how to start the shell with a script reloaded, refer to the *Connecting to a single node from the Mongo shell with a preloaded JavaScript* recipe in *Chapter 1, Installing and Starting the MongoDB Server*.

How to do it...

1. Load the required data in the collection using the `addTTLTestData2` method. Execute the following command on the Mongo shell:

    ```
    > addTTLTestData2()
    ```

2. Now, create the TTL index on the `ttlTest2` collection:

    ```
    > db.ttlTest2.ensureIndex({expiryDate :1}, {expireAfterSeconds:0})
    ```

3. Execute the following `find` query to view the three documents in the collection:

    ```
    > db.ttlTest2.find()
    ```

4. Now, after approximately 4, 5, and 7 minutes, see the documents with IDs 2, 1, and 3, respectively, getting deleted.

How it works...

Let's start by opening the `TTLData.js` file and seeing what must be going on in it. Our method for this recipe is `addTTLTestData2`. This method simply creates three documents in the `tllTest2` collection with `_id` of 1, 2, and 3, with their `exipryDate` field set to 5, 4, and 7 minutes, respectively, after the current time. Note that this field has a future date, unlike the date given in the previous recipe where it was a creation date.

Next, we will create an index:

```
> db.ttlTest2.ensureIndex({expiryDate :1}, {expireAfterSeconds:0})
```

This is different from the way we created the index for the previous recipe, where the `expireAfterSeconds` field of the object was set to a non-zero value. This is how the value of the `expireAfterSeconds` attribute is interpreted. If the value is non-zero, that is, the time in seconds elapsed after a base time, then the document will be deleted from the collection by Mongo. This base time is the value held in the field on which the index is created (`createTime`, as in the previous recipe). If this value is `0`, the date value on which the index is created (`expiryDate` in this case) will be the time when the document will expire.

To conclude, TTL indexes work well if you want to delete the document upon expiry. There are quite a few cases where we might want to move the document to an archive collection where the archived collection might be created based on, say, the year and month. In any such scenario, the TTL index is not helpful, and we might ourselves have to write an external job that does this work or reads the collection for a range of documents, adds them to the target collection, and deletes them from the source collection. JIRA (https://jira.mongodb.org/browse/SERVER-6895) is already open to address this issue. You might want to keep an eye on JIRA for further development on it.

There's more...

In this and the previous recipe, we looked at what TTL indexes are and how to use them. However, what if after creating a TTL index we want to modify it to change the value of the `expireAfterSeconds` value? It is possible using the `collMod` option. See more on this option in *Chapter 4, Administration*.

3
Programming Language Drivers

In this chapter, we will cover the following recipes:

- Installing PyMongo
- Executing query and insert operations using PyMongo
- Executing update and delete operations using PyMongo
- Aggregation in Mongo using PyMongo
- MapReduce in Mongo using PyMongo
- Executing query and insert operations using a Java client
- Executing update and delete operations using a Java client
- Aggregation in Mongo using a Java client
- MapReduce in Mongo using a Java client

Introduction

What we have seen so far using MongoDB is that we execute the majority of operations from the shell. The Mongo shell is a great tool for administrators to perform administrative tasks and for developers who would like to quickly test things by querying the data before coding the logic in the application. However, how do we write application code that will allow us to query, insert, update, and delete (among other things) the data in MongoDB? There has to be a library for the programming language in which we write our application.

Programming Language Drivers

We should be able to instantiate something or invoke methods from the program to perform some operations on the remote Mongo process. How will this happen unless there is a bridge that understands the protocol of communication with the remote server and is able to transmit the operation to execute over the wire we require on the MongoDB server process and get the result back to the client? This bridge, simply put, is called the driver, which is also referred to as a client library. Drivers form the backbone of Mongo's programming language interface. In the absence of drivers, it would have been the responsibility of the application to communicate with the MongoDB server using a protocol that the server understands. This would have been a lot of work not only to develop but also to test and maintain. Though the communication protocol is standard, there cannot be one implementation that works for all languages. A variety of programming languages need to have their own implementations that expose, more or less, the same sort of programming interface to all languages. The core concepts of client APIs that we will see in the chapter holds good for all languages.

Mongo has support for all the major programming languages and is supported by MongoDB, Inc. There is even a huge array of programming languages supported by the community. You might take a look at the various platforms supported by Mongo by visiting `http://docs.mongodb.org/ecosystem/drivers/community-supported-drivers/`.

To know more about the underlying protocol used by MongoDB or about communication between the client and the server, and to see what goes over the wire, refer to my blog at `http://amolnayak.blogspot.in/2014/09/mongodb-wire-protocol-analysis_14.html`.

Installing PyMongo

This recipe is about setting up PyMongo, which is the Python driver for MongoDB. In this recipe, we will demonstrate the installation of PyMongo on both the Windows and Linux platforms.

Getting ready

A simple single node is what we will need for the sanity testing of the driver, once the installation is complete. Refer to the *Single node installation of MongoDB* recipe in *Chapter 1, Installing and Starting the MongoDB Server*, to learn how to start the server. We will also require an Internet connection to download Python and PyMongo. Once these prerequisites are met, we are ready to begin.

The first step is to install Python on the computer if it is not already there. Visit `http://www.python.org/getit/`, download the latest version of Python for your platform, and install it. The steps for the installation Python are not covered in this recipe. However, before you proceed to the next section, Python should be available on the host operating system.

How to do it...

1. We will first set up PyMongo on the Windows platform. Visit https://pypi.python.org/pypi/pymongo and download the MS Windows Installer that is appropriate for the version of Python installed. My Python version is 2.7, and hence, this was the version downloaded.

2. Double-click on the downloaded installer and click on **Next**, as shown in the following screenshot:

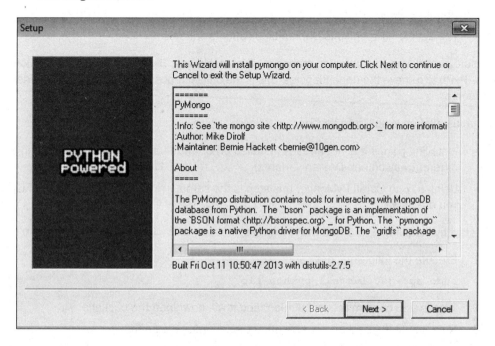

3. On clicking **Next**, if the right version of the Python installation is found, we can go ahead with the installation. With a couple more clicks on the **Next** button, we should have PyMongo installed.

Let's do a sanity test of the installation as follows:

1. From the command prompt, start the Python shell by typing in python as follows:

   `C:\Users\Amol>python`

   `Python 2.7.5 (default, May 15 2013, 22:43:36) [MSC v.1500 32 bit (Intel)] on win32`

   `Type "help", "copyright", "credits" or "license" for more information.`

2. We will then import PyMongo; this should happen without any error. If we don't see any import error, our installation has gone through successfully:

   ```
   >>> import pymongo
   >>>
   ```

In this section, we will see how to set up PyMongo on a Linux system. We will install it on a Debian flavor, Ubuntu. We will not use Ubuntu's **advanced packaging tool** (**apt**) to install PyMongo for a couple of reasons:

- Ubuntu's default repository might not have the latest release of the driver
- The apt tool is specific to Debian Linux and its variants

Therefore, we will use `pip`, a tool to manage Python packages. This tool uses **Python Package Index** (**PyPI**) to retrieve the dependencies; this is the official repository for third-party libraries in Python.

So, our installation is split into sections as follows:

- Installing `pip`, if it is not already installed on Ubuntu, using apt (it will be done in a different way on non-Debian variants)
- Using `pip` to install PyMongo; this step is the same, irrespective of the flavor of Linux you are using

Let's start by installing `pip` on Ubuntu as follows:

1. Execute the following command:

   ```
   sudo apt-get install python-pip
   ```

2. Type in `y` to confirm the installation, and it will download the package.
3. Now, install the package by executing the following command:

   ```
   amol@Amol-PC:~$ sudo apt-get install python-pip
   ```

4. Once the setup is complete, execute the following command from the shell to install PyMongo:

   ```
   $ pip install pymongo
   ```

5. This will install PyMongo. My Python version is the 2.7.x release. For the Python 3.x release, use `pymongo3` as the package name instead:

   ```
   $ pip install pymongo3
   ```

6. Once the installation of PyMongo is complete, we will do a quick sanity test. From the command prompt of the operating system, start the Python shell by typing in `python`:

   ```
   amol@Amol-PC:~$ python
   Python 2.7.3 (default, Apr 10 2013, 05:46:21)
   [GCC 4.6.3] on linux2
   Type "help", "copyright", "credits" or "license" for more information.
   ```

7. We will then import PyMongo; this will happen without any error:

   ```
   >>> import pymongo
   >>>
   ```

8. We are now done with the installation of the PyMongo setup.

There's more...

Installation of PyMongo is just a prerequisite for running Python code that can connect to Mongo to perform the operations. The next couple of recipes, *Executing query and insert operations using PyMongo* and *Executing update and delete operations using PyMongo*, are all about demonstrating these basic operations in Python programming using PyMongo to connect to the database and execute them.

Executing query and insert operations using PyMongo

This recipe is all about executing basic query and insert operations using PyMongo. This is similar to what we did with the Mongo shell earlier in the book.

Getting ready

To execute simple queries, we need to have a server up and running. A simple single node is what we will need. Refer to the *Single node installation of MongoDB* recipe in *Chapter 1, Installing and Starting the MongoDB Server*, to learn how to start the server. The data on which we will operate needs to be imported in the database. The steps to import the data are given in the *Creating test data* recipe in *Chapter 2, Command-line Operations and Indexes*. Python is expected to be installed on the host operating system and Mongo's client for python, PyMongo, needs to be installed. Look at the previous recipe to know how to install PyMongo for your host operating system. Also, in this recipe, we will execute insert operations and provide a write concern to use.

Programming Language Drivers

How to do it...

Let's start with some querying for Mongo from the Python shell. This will be identical to what we do from the Mongo shell, except that this is in the Python programming language as opposed to JavaScript that we have in the Mongo shell. We can use the basics that we will see here to develop large scale production systems that run on Python and use MongoDB as a data store.

Let's get started by first starting the Python shell from the operating system's command prompt. The following steps are independent of the host operating system:

1. Type in the following command in the shell, and the Python shell will start:

   ```
   $ python
   Python 2.7.5 (default, May 15 2013, 22:43:36) [MSC v.1500 32 bit
   (Intel)] on win32
   Type "help", "copyright", "credits" or "license" for more
   information.
   >>>
   ```

2. Then, import the `pymongo` package and create the client as follows:

   ```
   >>> import pymongo
   >>> client = pymongo.MongoClient('localhost', 27017)
   ```

   An alternative way to connect is as follows:

   ```
   >>> client = pymongo.MongoClient('mongodb://localhost:27017')
   ```

3. This works well too and achieves the same result. Now that we have the client, our next step is to get the database on which we will perform the operations. Now, unlike some programming languages where we have a `getDatabase()` method to get an instance of the database, we will get a reference to the database object on which we will perform the operations (`test` in this case). We will do this in the following way:

   ```
   >>> db = client.test
   ```

   Another alternative way is as follows:

   ```
   >>> db = client['test']
   ```

4. We will query the `postalCodes` collection. We will limit our results to 10 items as follows:

   ```
   >>> postCodes = db.postalCodes.find().limit(10)
   ```

5. Iterate over the results as follows. Watch out for the indentation of the print after the `for` statement. The following fragment should print 10 documents that are returned:

   ```
   >>> for postCode in postCodes:
     print 'City: ', postCode['city'], ', State: ',
   postCode['state'], ', Pin Code: ', postCode['pincode']
   ```

6. To find one document, execute the following command:

   ```
   >>> postCode = db.postalCodes.find_one()
   ```

7. Print the state and city of the returned result as follows:

   ```
   >>> print 'City: ', postCode['city'], ', State: ', postCode['state'], ', Pin Code: ', postCode['pincode']
   ```

8. Let's query the top 10 cities in the state of Gujarat sorted by the name of the city, and we will just select the city, state, and the pin code. Execute the following query from the Python shell:

   ```
   >>> cursor = db.postalCodes.find({'state':'Gujarat'}, {'_id':0, 'city':1, 'state':1, 'pincode':1}).sort('city', pymongo.ASCENDING).limit(10)
   ```

   The preceding cursor's results can be printed in the same way in which we printed the results in step 5.

9. Let's sort the data we query. We want to sort by the descending order of the state and then by the ascending order of the city. We will write the query as follows:

   ```
   >>> city = db.postalCodes.find().sort([('state', pymongo.DESCENDING),('city',pymongo.ASCENDING)]).limit(5)
   ```

10. Iterate through this cursor; this should print out five results on the console. Refer to step 5 for how we iterate over a cursor returned to print the results.

11. So, we played a bit to find documents and covered basic operations from Python as far as querying MongoDB is concerned. Now, let's see a bit about the `insert` operation. We will use a `test` collection to perform these operations and not disturb our postal codes test data. We will use a `pymongoTest` collection for this purpose and add documents in a loop to it as follows:

    ```
    >>> for i in range(1, 21):
            db.pymongoTest.insert({'i':i})
    ```

12. The `insert` operation can take a list of dictionary objects and perform a bulk insert. So now, something like the following `insert` query is perfectly valid:

    ```
    >>> db.pythonTest.insert([{'name':'John'}, {'name':'Mark'}])
    ```

    Any guesses on the return value? In the case of a single document insert, the return value is the value of `_id` for the newly created document. In this case, it is a list of IDs.

13. Let's execute an `insert` query again, this time, with a write concern provided. Execute the following write concern with w = 1 and j = True:

    ```
    >>> db.pymongoTest.insert({'name': 'Jones'}, w = 1, j = True)
    ```

How it works...

We instantiated the client and then got the reference to the object that will be used to access the database on which we wish to perform operations in step 3. There are a couple of ways to get this reference. The first option (`db = client.test`) is more convenient, unless your database name has a special character, such as a hyphen (-). For example, if the name is `db-test`, we would have no option other than to use the `[]` operator to access the database. Using either of the alternatives, we now have an object for the `test` database in the `db` variable. After we got the client and the `db` instance in Python, we queried to find the top 10 documents in the natural order from the collection in step 4. The syntax is exactly identical to how this query would have been executed from the shell. Step 5 simply printed out the results, 10 of them in this case. Generally, if you need instant help on a particular class using the class name or an instance of this class from the Python interpreter, simply execute `dir(<class_name>)` or `dir(<object of a class>)`; which gives a listing of the attributes and functions defined in the module passed. For example, `dir('pymongo.MongoClient')` or `dir(client)`, where `client` is the variable that holds the reference to an instance of `pymongo.MongoClient`, can be used to get the listing of all the supported attributes and functions. The `help` function is more informative and prints out the module's documentation, which is a great source of reference just in case you need instant help. Try typing in `help('pymongo.MongoClient')` or `help(client)`.

In steps 4 and 5, we queried the `postalCodes` collection, limited the result to the top 10 results, and printed them. The returned object is of type `pymongo.cursor.Cursor` class. The next step got just one document from the collection using the `find_one()` function. This is synonymous to the `findOne()` method on the collection invoked from the shell. The value returned by this function is an inbuilt `dict` object.

In step 8, we executed another find to query the data. However, this time around, we passed two parameters to it. The first one was the query, which looked similar to how we execute from the Mongo shell. However, the type of the parameter in Python is `dict`. The second parameter was another object of type `dict`. This dictionary is used to provide the fields to be returned in the result. A value `1` for a field indicates that the value is to be selected and returned in the result. This is synonymous to `select` in the relational database, with a few sets of columns provided explicitly to be selected. The `_id` field is selected by default, unless it is explicitly set to 0 in the selector `dict` object. The selector provided here is `{'_id':0, 'city':1, 'state':1, 'pincode':1}`, which selects the city, state, and pin code and suppresses the `_id` field. We have the `sort` method too. This method has two formats: `sort(sort_field, sort_direction)` and `sort([(sort_field, sort_direction)...(sort_field, sort_direction)])`.

The first one is used when we want to sort by one field only. The second representation accepts a list of pairs of sort fields and sort directions and is used when we want to sort by multiple fields. We used the first format in the query in step 8 and the second format in our query in step 9, as we sorted first by state name and then by city.

If we look at the way we invoked sort, it was invoked on the `cursor` instance. Similarly, the `limit` function was also on the `Cursor` class. The evaluation is lazy and is deferred until the iteration is performed to retrieve the results from the cursor. Until that point, the `cursor` object is not evaluated on the server.

In step 12, we inserted a document 20 times in a collection. Each insert, as we see in the Python shell, will return a generated `_id` field. In terms of the syntax of insert, it is exactly identical to the operation we perform from the shell. The parameter passed for the `insert` operation is again an object of type `dict`.

In step 13, we passed a list of documents to insert in the collection. This inserts multiple documents in one call to the server; this is a bulk insert. The return value in this case is a list of IDs, one for each document inserted and in the same order as passed in the input list. However, as MongoDB doesn't support transactions, all inserts will be independent of each other, and a failure of one insert doesn't automatically roll back the entire operation.

Adding to the functionality to insert multiple documents demanded another parameter for the behavior. When one of the inserts in the given list fails, should the remaining inserts continue or should the insertion stop as soon as the first error is encountered? The name of the parameter to control this behavior is `continue_on_error`, and its default value is `False`, that is, stop as soon as the first error is encountered. If this value is `True` and multiple errors occur during insertion, only the latest error will be available. Hence, the default option is `False`, as the value is sensible. Let's take a look at a couple of examples. In the Python shell, execute the following commands:

```
>>> db.contOnError.drop()
>>> db.contOnError.insert([{'_id':1}, {'_id':1}, {'_id':2}, {'_id':2}])
>>> db.contOnError.count()
```

The count we will get is 1, which is for the first document with the `_id` field as 1. The moment another document with the same value of the `_id` field is found, 1 in this case, an error is thrown, and the bulk insert stops. Now, execute the following `insert` operation:

```
>>> db.contOnError.drop()
>>> db.contOnError.insert([{'_id':1}, {'_id':1}, {'_id':2}, {'_id':2}], continue_on_error=True)
>>> db.contOnError.count()
```

Here, we passed an additional parameter, `continue_on_error`, whose value is `True`. As a result of this parameter, the `insert` operation will continue with the next document even if an intermediate `insert` operation failed. The second insert with `_id:1` fails; yet, the next insert goes through before another insert with `_id:2` fails (as one document with this `_id` is already present). Also, the error reported is for the last failure, the one with `_id:2` in this case.

Another parameter is `check_keys`, which checks for key names that start with $ and the existence of . in the key. If one is found, it will raise `bson.errors.InvalidDocument`. Thus, the following insert operation will fail:

```
>>> db.pymongoTest.insert({'a.b':1})
```

By default, the check will take place, unless you explicitly disable it by setting the value of this parameter to `False`. Thus, the following query will pass and return an object ID of the inserted document:

```
>>> db.pymongoTest.insert({'a.b':1}, check_keys=False)
```

Step 13 executed the `insert` operation but provided a write operation to be used for the insert to be executed.

See also

- The *Executing update and delete operations using PyMongo* recipe

Executing update and delete operations using PyMongo

In the previous recipe, we saw how to execute `find` and `insert` operations in MongoDB using PyMongo. In this recipe, we will see how updates and deletions work from Python. We will also see what atomic find and update/delete is and how to execute these operations. We will then conclude by revisiting `find` operations and look at some interesting functions of the `cursor` object.

Getting ready

If you have already seen and completed the previous recipe, you are all set to go. If not, it is recommended that you first complete the previous recipe before going ahead with this recipe.

Before we get started, let's define a small function that iterates through the cursor and shows the results of a cursor on the console. We will use this function whenever we want to display the results of a query on the `pymongoTests` collection. The following is the function's body:

```
>>> def showResults(cursor):
        if cursor.count() != 0:
            for e in cursor:
                print e
        else:
            print 'No documents found'
```

Also, refer to steps 1 and 2 in the previous recipe to learn how to create a connection to the MongoDB server and create the `db` object used to perform the CRUD operation on this database. Also, refer to step 11 in the previous recipe to learn how to insert the required test data in the `pymongoTest` collection. You might confirm the data in this collection by executing the following command from the Python shell once the data is present:

```
>>> showResults(db.pymongoTest.find())
```

For a part of the recipe, one is also expected to know and start a replica set instance. Refer to the *Starting multiple instances as part of a replica set* and *Connecting to the replica set from the shell to query and insert data* recipes in *Chapter 1, Installing and Starting the MongoDB Server*.

How to do it...

1. We will set a field named `gtTen`, specifying with a Boolean value `True` if the field `i` has a value greater than 10. Let's execute the following `update` command:

   ```
   >>> db.pymongoTest.update({'i':{'$gt':10}},
   {'$set':{'gtTen':True}})
   {u'updatedExisting': True, u'connectionId': 8, u'ok': 1.0, u'err': None, u'n': 1}
   ```

2. Query the collection and view its data by executing the following command, and check the data that got updated:

   ```
   >>> showResults(db.pymongoTest.find())
   ```

3. The results displayed confirm that only one document got updated. We will now execute the same update again but, this time around, we will update all the documents that match the provided query. Execute the following `update` operation from the Python shell. Note that this update is identical to the one we performed in step 1, except for the additional parameter called `multi` whose value is given as `True`. Also, note the value of `n` in the response; it is `10` this time:

   ```
   >>> db.pymongoTest.update({'i':{'$gt':10}},{'$set':{'gtTen':True}}, multi=True)
   {u'updatedExisting': True, u'connectionId': 8, u'ok': 1.0, u'err': None, u'n': 10}
   ```

4. Execute the operation we performed in step 2 again to view the contents in the `pymongoTest` collection and verify the documents updated.

5. Let's take a look at how `upsert` operations can be performed. Upserts are updates plus inserts. They update a document if one exists, just as an update will do; otherwise, it will insert a new document. Let's take a look at an example. Consider the following command on a document that doesn't exist in the collection:

   ```
   >>> db.pymongoTest.update({'i':21},{'$set':{'gtTen':True}})
   ```

6. The update here will not update anything and will return the number of updated documents as 0. However, let's consider that we want to update a document if it exists or insert a new document and apply the update on it atomically and then perform an `upsert` operation. In this case, the `upsert` operation is executed as follows (note the response that mentions `upsert`, `ObjectId` of the newly inserted document, and the `updatedExisting` value, which is `False`):

   ```
   >>> db.pymongoTest.update({'i':21},{'$set':{'gtTen':True}}, upsert=True)
   {u'ok': 1.0, u'upserted': ObjectId('52a8b2f47a809beb067ecd8a'), u'err': None, u'connectionId': 8, u'n': 1, u'updatedExisting': False}
   ```

7. Let's see how to delete documents from the collection using the `remove` method:

   ```
   >>> db.pymongoTest.remove({'i':21})
   {u'connectionId': 8, u'ok': 1.0, u'err': None, u'n': 1}
   ```

8. If we look at the value of `n` in the preceding response, we will see that it is 1. This means that one document got removed. There is another way to remove the document by `_id`. Let's insert one document in the collection and later remove it. Insert the document as follows:

   ```
   >>> db.pymongoTest.insert({'i':23, '_id':23})
   ```

9. Now, remove this document from the collection as follows:

   ```
   >>> db.pymongoTest.remove(23)
   {u'connectionId': 8, u'ok': 1.0, u'err': None, u'n': 1}
   ```

10. We will look at the `find` and `modify` operations now. We can look at this operation as a way to find a document and then update/remove it; both of these operations are performed atomically. Once the operation is performed, the document returned is either the one before or after the `update` operation was done (in the case of `remove`, there will be no document after the operation). In the absence of this operation, we cannot guarantee an atomic find, update the document, and return the resulting document before/after the update in scenarios where multiple client connections could be performing similar operations on the same document. The following is an example of how to perform these `find` and `modify` operations in Python:

    ```
    >>> db.pymongoTest.find_and_modify({'i':20}, {'$set':{'inWords':'Twenty'}})
    {u'i': 20, u'gtTen': True, u'_id': ObjectId('52a8a1eb072f651578ed98b2')}
    ```

    The preceding result shows us that the resulting document returned is the one before the update was applied.

11. Execute the following `find` operation to query and view the document that we updated in the previous step. The resulting document will contain the newly added `inWords` field:

    ```
    >>> db.pymongoTest.find_one({'i':20})
    {u'i': 20, u'_id': ObjectId('52aa0cfe072f651578ed98b7'),
    u'inWords': u'Twenty'}
    ```

12. We will execute the `find` and `modify` operations again but, this time around, we will return the updated document rather than the document before the update, which we saw in step 9. Execute the following command from the Python shell:

    ```
    >>> db.pymongoTest.find_and_modify({'i':19}, {'$set':{'inWords':'N
    ineteen'}}, new=True)
    {u'i': 19, u'gtTen': True, u'_id': ObjectId('52a8a1eb072f651578ed9
    8b1'), u'inWords': u'Nineteen'}
    ```

13. We saw how to query using PyMongo in the previous recipe. Here, we will continue with the query operation. We saw how the `sort` and `limit` functions were chained to the `find` operation. The prototype of the call on the `postalCodes` collection is as follows:

    ```
    db.postalCode.find(..).limit(..).sort(..)
    ```

14. There is an alternate way that achieves the same result as the one achieved earlier. Execute the following query in the Python shell to achieve the same result:

    ```
    >>> cursor = db.postalCodes.find({'state':'Gujarat'}, {'_id':0,
    'city':1, 'state':1, 'pincode':1}, limit=10, sort=[('city',
    pymongo.ASCENDING)])
    ```

15. Print the preceding cursor using the `showResult` function that is already defined.

16. To restrict a full table scan on the collection by queries without indexes, there is a parameter called `max_scan`, which takes an integer value. This value of the `max_scan` parameter ensures that a query doesn't scan more than the value provided. For instance, the following query ensures that no more than 50 documents are scanned to get the results. Again, use the `showResults` function to display the results in the cursor:

    ```
    >>> showResults(db.postalCodes.find({'state':'Andhra Pradesh'},
    max_scan=50))
    ```

Programming Language Drivers

How it works...

Let's take a look at what we did in this recipe. We started by updating the documents in a collection in step 1. The update, however, updated only the first matching document by default and the rest of the matching documents were not updated. In step 2, we added a parameter called `multi` with a value `True` to update multiple documents as part of the same `update` operation. Note that all these documents are not updated atomically as part of one transaction. If we look at the update done from the Python shell, we will see a striking resemblance to what we would have done from the Mongo shell. If we want to name the arguments of the `update` operation, the names of the parameter are called `spec` and `document`, which are for the document provided as a query to be used to select the documents and to update documents respectively. For instance, the following `update` operation is valid:

```
>>> db.pymongoTest.update(spec={'i':{'$gt':10}},document=
{'$set':{'gtTen':True}})
```

There are some more arguments that an `update` function takes, with most of them carrying the same meaning as the `insert` function we saw in the previous recipe. These parameters are `w`, `wtimeout`, `j`, `fsync`, and `check_keys`. Refer to the previous recipe for the explanation given for these parameters used with the `insert` function.

In step 6, we did an upsert (update plus insert). All we had was an additional `upsert` parameter with the value as `True`. However, what exactly happens in the case of an upsert? Mongo tries to update the document that matches the provided condition; if it finds one, this will have been a regular update. However, in this case (upsert in step 6), the document was not found. The server inserted the document given as `spec` (the first parameter) in the collection and then applied an `update` operation on it with both these operations taking place atomically.

In steps 7 and 8, we saw the `remove` operation. The first variant accepted a query and all the matching documents were removed. The second variant, in step 8, accepted one integer, which is the value of the `_id` field to be deleted. This variant is useful whenever we plan to delete by the `_id` field's value. Similar to `update`, the `remove` function too accepts other parameters for the write concern. The `w`, `wtimeout`, `j`, and `fsync` parameters have meanings similar to what we discussed in the previous recipe when we inserted the documents. Refer to the previous recipe for a detailed description of these parameters. The call to the `remove` method on the collection without any parameter will remove all the documents in the collection.

In steps 10 to 12, we executed the `find` and `modify` operations. Information on these operations is provided in the previous section. What we didn't see is that this operation can also be used to find and remove documents from the collection. An additional parameter called `remove` needs to be added with the value as `True`. In the following operation, we will remove the document with `_id` equals 31 and return the document before deleting it:

```
>>> db.pymongoTest.find_and_modify(query={'_id':31}, remove=True)
```

Note that, with the `remove` option provided, the parameter named `new` is not supported, as there is nothing to return after the document is deleted.

All the operations we saw in this recipe were for the clients connected to a standalone instance. If, however, you are connected to a replica set, the client is instantiated in a different way. Also, we are aware of the fact that, by default, we are not allowed to query the secondary nodes for data. We need to explicitly execute `rs.slaveOk()` from the Mongo shell connected to a secondary node to query it. This is done in a similar way from a Python client as well. If we are connected to a secondary node, we cannot query it by default, but the way in which we specify that we are ok to query on a secondary node is slightly different. There is a parameter called `slave_okay` to let us query from the secondary node whose value is `False` by default; if the value is `True`, the query will go through successfully and return results from a secondary node. If the parameter is not set to `True`, querying the secondary node will throw an exception that states that the node queried is not a master. For instance, if our client is connected to a secondary instance and we want to query it based on the name of the state, we will execute the following query:

```
>>> cursor = db.postalCodes.find({'state':'Maharashtra'}, slave_ok=True)
```

We will get the cursor for the results successfully if the collection does indeed have documents with the name of the state, `Maharashtra`.

Another parameter that is better left untouched and has a sensible default is called `timeout`, and its value by default is `True`. Note that this value is not a number for some sort of timeout but a Boolean value. If the value is `True`, the cursor opened by a query on the server will be auto-closed after 10 minutes of inactivity on it. Let's say, it is a sort of a garbage collection of the server-side resources. However, if this is set to `False`, it is no longer the responsibility of the server to clean it up, but the responsibility of the client to close it.

Another parameter called `tailable` is used to denote that the cursor returned by `find` is a tailable cursor. Explaining what tailable cursors are and giving more details is not in the scope of this recipe; this is explained in the *Creating and tailing capped collection cursors in MongoDB recipe* in *Chapter 5, Advanced Operations*.

So far in the recipe, we connected to a single node using `pymongo.MongoClient`. However, we cannot use the same class to connect to a replica set because of the following reasons:

- We will just be connected to one instance
- To allow us to perform write operations, we will have to connect to the primary instance
- If the primary instance goes down, there has to be an automatic failover to the new primary instance

Programming Language Drivers

Therefore, to connect to a replica set and address the preceding three points, we will use `pymongo.MongoReplicaSetClient`. The following is the way in which we can initiate the client:

```
>>> client = pymongo.MongoReplicaSetClient('mongodb://localhost:27000', replicaSet='replSetTest')
>>>
```

As we can see, we just provided one host from the replica set and the name of the replica set we used when starting it. The client will automatically discover the remaining hosts from the replica set configuration. The host name(s) that we provided is known as the seed list, using which we can provide multiple instances in the replica set. The name of the parameter that gives the host names is `hosts_or_uri`.

However, what about read preferences and how do we specify them? There are some more parameters that we will need to look at while initiating the client.

```
>>> from pymongo.read_preferences import ReadPreference
>>> from pymongo import MongoReplicaSetClient
>>> client = MongoReplicaSetClient('mongodb://localhost:27000', replicaSet='replSetTest', read_preference=ReadPreference.NEAREST)
>>> client.read_preference
4
```

The preceding steps initialized a replica set client with a read preference `NEAREST`. There is an additional parameter, `secondary_acceptable_latency_ms`, which gives the time in milliseconds. Now, this time will be used by the client to consider a member of the replica set as a contender for selection when the read preference `NEAREST` is specified. A minimum latency is first computed for all the replica set instances from the driver, and all the instances with a latency no more than the provided value will be added to the contender instances' list for selection as the nearest instance to the driver. There was a fairly long discussion on this behavior in the read preference recipe, and some code snippets from a Java client were used to explain the internals. The default value for this parameter is 15 milliseconds.

As we know, read preference can be provided at the client level, at the database level that gets inherited from the client, and also at the cursor level. By default, `read_preference` for a client initialized without an explicit read preference is `PRIMARY` (with the value 0). However, if we now get the database object from the client initialized earlier, the read preference will be `NEAREST` (with the value 4).

```
>>> db = client.test
>>> db.read_preference
4
>>>
```

Setting the read preference is as simple as executing the following command:

```
>>> db.read_preference = ReadPreference.PRIMARY_PREFERRED
```

Again, as the read preference gets inherited from the client to the database object, it gets inherited from the database object to the collection object, and it will be used as the default value for all the queries executed against that collection, unless read preference is specified explicitly in the `find` operation.

Thus, `db.pymongoTest.find()` will have a cursor, which uses the read preference as `PRIMARY_PREFERRED` (we just set it earlier to `PRIMARY_PREFERRED` at the database-object level) whereas `db.pymongoTest.find(read_preference=ReadPreference.NEAREST)` will use the read preference as `NEAREST`.

We will now wrap up the basic operations from a Python driver by trying to do some common operations that we do from the Mongo shell, such as getting all the database names, getting a list of collections in a database, and creating an index on a collection.

From the shell, we will execute `show dbs` to show all the database names in the Mongo instance that is connected. From the Python client, we will execute the following command on the client instance:

```
>>> client.database_names()
[u'local', u'test']
```

Similarly, to see the list of collections, we will type `show collections` in the Mongo shell. In Python, all that we will do on the database object is as follows:

```
>>> db.collection_names()
[u'system.indexes', u'writeConcernTest', u'pymongoTest']
```

Now, for index operations, we will first see what indexes are present in the `pymongoTest` collection. Execute the following command from the Python shell to view the indexes on a collection:

```
>>> db.pymongoTest.index_information()
{u'_id_': {u'key': [(u'_id', 1)], u'v': 1}}
```

We now will create an index on key x, which is sorted in ascending order on the `pymongoTest` collection as follows:

```
>>> db.pymongoTest.ensure_index([('x',pymongo.ASCENDING)])
u'x_1'
```

We can again list the indexes as follows to confirm the creation of the index:

```
>>> db.pymongoTest.index_information()
{u'_id_': {u'key': [(u'_id', 1)], u'v': 1}, u'x_1': {u'key': [(u'x', 1)], u'v':1}}
```

We can see that the index got created. Generally speaking, the format of the `ensure_index` method is as follows:

```
>>> db.<collection name>.ensure_index([(<field name 1>,<order of field 1>)…..(<field name n >,<order of  field n>)])
```

Aggregation in Mongo using PyMongo

We already saw PyMongo using Python's client interface for MongoDB in the *Executing query and insert operations using PyMongo* and *Executing update and delete operations using PyMongo* recipes. In this recipe, we will use the postal code collection and run an aggregation example using PyMongo. The intention of this recipe is not to explain aggregation but to show how aggregation can be implemented using PyMongo. In this recipe, we will aggregate the data based on the state names and get the top five state names by the number of documents they appear in. We will make use of the `$project`, `$group`, `$sort`, and `$limit` operators for the process.

Getting ready

To execute the aggregation operations, we need to have a server up and running. A simple single node is what we will need. Refer to the *Single node installation of MongoDB* recipe in *Chapter 1, Installing and Starting the MongoDB Server*, to learn how to start the server. The data on which we will operate needs to be imported in the database. The steps to import the data are given in the *Creating test data* recipe in *Chapter 2, Command-line Operations and Indexes*. Python and PyMongo are expected to be installed. Look at the *Installing PyMongo* recipe to know how to install PyMongo for your host operating system. Since this is a way to implement aggregation in Python, that the reader is expected to be aware of the aggregation framework on MongoDB.

How to do it...

Let's take a look at the steps in detail:

1. Open the Python terminal by typing the following command:

   ```
   $ python
   ```

2. Once the Python shell opens, import PyMongo as follows:

   ```
   >>> import pymongo
   ```

3. Create an instance of `MongoClient` as follows:

   ```
   >>> client = pymongo.MongoClient('mongodb://localhost:27017')
   ```

4. Get the `test` database's object as follows:

   ```
   >>> db = client.test
   ```

5. Now, we will execute the aggregation operation on the `postalCodes` collection as follows:

   ```
   result = db.postalCodes.aggregate(
     [
       {'$project':{'state':1, '_id':0}},
       {'$group':{'_id':'$state', 'count':{'$sum':1}}},
       {'$sort':{'count':-1}},
       {'$limit':5}
     ]
   )
   ```

6. Type the following command to view the results:

   ```
   >>> result['result']
   ```

How it works...

The steps are pretty straightforward. We connected to the database that runs on the localhost and created a database object. The aggregation operation we invoked on the collection using the `aggregate` function is very similar to how we will invoke aggregation from the shell. The object in the return value, `result`, is an object of type `dict`; it has two keys of interest. One of the keys is called `ok`, whose value will be `1` if the aggregation operation executed successfully. The other key is called `result` and its type is a list. In our case, it will contain five documents that contain the name of the state and the count of the number of their occurrences.

MapReduce in Mongo using PyMongo

In the previous recipe, we saw how to execute aggregation operations in Mongo using PyMongo. In this recipe, we will work on the same use case as we did for the aggregation operation, but using MapReduce. The intent is to aggregate the data based on state names and get the top five state names by the number of documents they appear in.

Programming language drivers provide us with an interface to let us invoke MapReduce jobs written in JavaScript on the server.

Getting ready

To execute the MapReduce operations, we need to have a server up and running. A simple single node is what we will need. Refer to the *Single node installation of MongoDB* recipe in *Chapter 1, Installing and Starting the MongoDB Server*, to learn how to start the server. The data on which we will be operating needs to be imported in the database. The steps to import the data are given in the *Creating test data* recipe in *Chapter 2, Command-line Operations and Indexes*. Python is expected to be installed on the host operating system, and PyMongo also needs to be installed. Take a look at the *Installing PyMongo* recipe to know how to install PyMongo for your host operating system.

How to do it...

Let's take a look at the steps in detail:

1. Open the Python terminal by typing in the following command:

   ```
   >>> python
   ```

2. Once the Python shell opens, import the bson package as follows:

   ```
   >>> import bson
   ```

3. Now, import the pymongo package as follows:

   ```
   >>> import pymongo
   ```

4. Create an instance of MongoClient as follows:

   ```
   >>> client = pymongo.MongoClient('mongodb://localhost:27017')
   ```

5. Get the test database's object as follows:

   ```
   >>> db = client.test
   ```

6. Write the mapper function as follows:

   ```
   >>> map = bson.Code('''function() {emit(this.state, 1)}''')
   ```

7. Write the reduce function as follows:

   ```
   >>> reduce = bson.Code('''function(key, values){return Array.sum(values)}''')
   ```

8. Invoke MapReduce as follows (note that the result will be sent to the pymr_out collection):

   ```
   >>> db.postalCodes.map_reduce(map=map, reduce=reduce, out='pymr_out')
   ```

9. Verify the result as follows:

```
>>> c = db.pymr_out.find(sort=[('value', pymongo.DESCENDING)], limit=5)
>>> for elem in c:
...     print elem
...
{u'_id': u'Maharashtra', u'value': 6446.0}
{u'_id': u'Kerala', u'value': 4684.0}
{u'_id': u'Tamil Nadu', u'value': 3784.0}
{u'_id': u'Andhra Pradesh', u'value': 3550.0}
{u'_id': u'Karnataka', u'value': 3204.0}
>>>
```

How it works...

Apart from the regular import for PyMongo, here we imported the `bson` package too. This is where we have the `Code` class that we use for writing the JavaScript `map` and `reduce` functions. It is instantiated by passing the JavaScript function body as a constructor argument.

Once two instances of the `Code` class are instantiated, one for map and one for reduce, all we need to do is invoke the `map_reduce` function on the collection. In this case, we passed three parameters: the two `Code` instances for the `map` and `reduce` functions with parameter names `map` and `reduce`, respectively, and one string value, used to provide the name of the output collection into which the results are written.

We won't be explaining the MapReduce JavaScript function here, but it is pretty simple; all it does is emit keys as the names of the states and values, which is the number of times the particular state name occurs. This resulting document with the key used, the state's name as the `_id` field, and another field called `value`, which is the sum of the times the particular state's name given in the `_id` field appeared in the collection, are added to the output collection, `pymr_out` in this case. For example, in the entire collection, the state, Maharashtra, appeared 6446 times. Thus, the document for the state of Maharashtra is `{u'_id': u'Maharashtra', u'value': 6446.0}`. To confirm whether this is a true value or not, you can execute the following query from the Mongo shell and see that the result is indeed 6446:

```
> db.postalCodes.count({state:'Maharashtra'})
6446
```

We are still not done as the requirement is to find the top five states by their occurrence in the collection. We still have just the states and their occurrences, so the final step is to sort the documents by the value field, which is the number of times the state's name occurred in the descending order, and limit the result to five documents.

See also

> ▶ Chapter 8, *Integration with Hadoop,* for different recipes on executing MapReduce jobs on MongoDB using the Hadoop connector, which allows us to write the `map` and `reduce` functions in languages such as Java and Python

Executing query and insert operations using a Java client

In this recipe, we will look at executing the query and insert operations using a Java client for MongoDB. Unlike the Python programming language, Java code snippets cannot be executed from an interactive interpreter. Thus, we will have some unit test cases already implemented; their relevant code snippets will be shown and explained.

Getting ready

For this recipe, we will start a standalone instance. Refer to the *Single node installation of MongoDB* recipe in *Chapter 1, Installing and Starting the MongoDB Server,* to learn how to start the server.

The next step is to download the `mongo-cookbook-javadriver` Java project from the book's website. This recipe uses a JUnit test case to test various features of the Java client. In this whole process, we will make use of some of the most common API calls and, thus, learn to use them.

How to do it...

To execute the test case, one can either import the project in an IDE such as Eclipse and execute the test case, or execute the test case from the command prompt using Maven.

The test case we are going to execute for this recipe is `com.packtpub.mongo.cookbook.MongoDriverQueryAndInsertTest`.

If you are using an IDE, open this test class and execute it as a JUnit test case. If you are planning to use Maven to execute this test case, go to the command prompt, change the directory to the root of the project, and execute the following command to execute this single test case:

```
$ mvn -Dtest=com.packtpub.mongo.cookbook.MongoDriverQueryAndInsertTest test
```

Everything should execute fine, and the test case should succeed if the Java SDK and Maven are properly set up and the MongoDB server is up and running and listening to port `27017` for incoming connections.

How it works...

We will now open the test class we executed and see some of the important API calls in the test method. The super class of our test class is `com.packtpub.mongo.cookbook.AbstractMongoTest`.

We will start by looking at the `getClient` method in this class. The client instance that gets created is an instance of type `com.mongodb.MongoClient`. There are several overloaded constructors for this class; however, we will use the following constructor to instantiate the client:

```
MongoClient client = new MongoClient("localhost:27017");
```

Another method to look at is `getJavaDriverTestDatabase` in the same abstract class that gets us the database instance. This instance is synonymous to the implicit variable db in the shell. Here, in Java, this class is an instance of type `com.mongodb.DB`. We will get an instance of this DB class by invoking the `getDB()` method on the client instance. In our case, we want the DB instance for the `javaDriverTest` database, which is as follows:

```
getClient().getDB("javaDriverTest");
```

Once we get the instance of `com.mongodb.DB`, we will use it to get the instance of `com.mongodb.DBCollection`, which will be used to perform various operations, `find` and `insert` in our case, on the collection. The `getJavaTestCollection` method in the abstract test class returns one instance of `DBCollection`. We will get an instance of `DBCollection` class for the `javaTest` collection by invoking the `getCollection()` method on `com.mongodb.DB` as follows:

```
getJavaDriverTestDatabase().getCollection("javaTest")
```

Once we get an instance of `DBCollection`, we will be ready to perform operations on it. In the scope of this recipe, it is limited to `find` and `insert` operations.

Now, we will open the main test case class `com.packtpub.mongo.cookbook.MongoDriverQueryAndInsertTest`. Open this class in an IDE or text editor. We will look at the methods of this class. The first method that we will look at is `findOneDocument`. Here, the line of our interest is `collection.findOne(new BasicDBObject("_id", 3));` this queries the document with the value of _id as 3

This method returns an instance of `com.mongodb.DBObject`, which is a key-value map that returns the fields of a document as a key and the value as the value of this corresponding key. For instance, to get the value of _id from the returned `DBObject` instance, we will invoke `result.get("_id")` on the returned result.

Programming Language Drivers

Our next method to inspect is `getDocumentsFromTestCollection`. This test case executes a `find` operation on the collection and gets all the documents in it. The `collection.find()` call executes the `find` operation on the DBCollection's instance. The return value of the `find` operation is `com.mongodb.DBCursor`. An important point to note is that invoking the `find` operation doesn't itself execute the query but just returns the DBCursor's instance. This is an inexpensive operation and doesn't consume server-side resources. The actual query gets executed on the server side only when the `hasNext` or `next` method is invoked on the `DBCursor` instance. The `hasNext()` method is used to check if there are more results, and the `next()` method is used to navigate to the next `DBObject` in the result. An example usage of the `DBCursor` instance returned to iterate through the results is as follows:

```
while(cursor.hasNext()) {
  DBObject object = cursor.next();
  //Some operation on the returned object to get the fields and
  //values in the document
}
```

We will now look at two methods: `withLimitAndSkip` and `withQueryProjectionAndSort`. These methods show us how to sort, limit the number of results, and skip the number of initial results. As we can see in the following code snippet, the methods sort, limit, skip, and chain to each other.

```
DBCursor cursor = collection
    .find(null)
    .sort(new BasicDBObject("_id", -1))
    .limit(2)
    .skip(1);
```

All these methods return an instance of `DBCursor` itself; this allows us to chain the calls. These methods are defined in the `DBCursor` class, which changes certain states according to the operation they perform in the instance and has `return this` at the end of the method to return the same instance.

Remember that the actual operation is invoked on the server only upon invoking the `hasNext` or `next` method on `DBCursor`. Invoking any method such as `sort`, `limit`, and `skip` after the execution of the query on the server will throw `java.lang.IllegalStateException`.

We used two variants of the `find` method: one that accepts one parameter for the query to be executed and another one that has two parameters; the first one is for the query and the second one is another `DBObject`, which is used for the projection that will return only a selected set of fields from the document in the result.

The following query, for instance, from the `withQueryProjectionAndSort` method of the test case, selects all the documents, as the first argument is null, and the returned `DBCursor` instance will have documents that contain just one field called `value`:

```
DBCursor cursor = collection
```

```
    .find(null, new BasicDBObject("value", 1).append("_id", 0))
    .sort(new BasicDBObject("_id", 1));
```

The `_id` field is to be explicitly set to 0; otherwise, it will be returned by default.

Finally, we will look at two more methods in the test case: `insertDataTest` and `insertTestDataWithWriteConcern`. We will use a couple of variants of the `insert` method in these two methods. All the `insert` methods are invoked on the `DBCollection` instance and return the `com.mongodb.WriteResult` instance. The result can be used to get the error that occurred during the write operation by invoking the `getLastError()` method, to get the number of documents inserted using the `getN()` method, and get the write concern of the operation among the small number of operations. Refer to the Javadoc of the MongoDB API at https://api.mongodb.org/java/current/ for more details on the methods. The two `insert` operations that we performed are as follows:

```
collection.insert(new BasicDBObject("value", "Hello World"));
collection.insert(new BasicDBObject("value", "Hello World"),
    WriteConcern.JOURNALED);
```

Both of these accept a `DBObject` instance for the document to be inserted as the first parameter. The second method allows us to provide the write concern to be used for the write operation. There are `insert` methods in the `DBCollection` class that allows bulk insert too. Refer to the Javadoc at https://api.mongodb.org/java/current/ for more details on various overloaded versions of the `insert` method.

Executing update and delete operations using a Java client

In the previous recipe, we saw how to execute the `find` and `insert` operations in MongoDB using a Java client. In this recipe, we will see how the `update` and `delete` operations work from a Java client.

Getting ready

For this recipe, we will start a standalone instance. Refer to the *Single node installation of MongoDB* recipe in *Chapter 1, Installing and Starting the MongoDB Server*, to learn how to start the server.

The next step is to download the Java project `mongo-cookbook-javadriver` from the book's website. This recipe uses a JUnit test case to test out various features of the Java client. In this whole process, we will make use of some of the most common API calls and, thus, learn how to use them.

How to do it...

To execute the test case, one can either import the project in an IDE such as Eclipse and execute the test case or execute the test case from the command prompt using Maven.

The test case we are going to execute for this recipe is `com.packtpub.mongo.cookbook.MongoDriverUpdateAndDeleteTest`.

If you are using an IDE, open this test class and execute it as a JUnit test case. If you plan to use Maven to execute this test case, go to the command prompt, change the directory to the root of the project, and execute the following command to execute this single test case:

```
$ mvn -Dtest=com.packtpub.mongo.cookbook.MongoDriverUpdateAndDeleteTest test
```

Everything should execute fine if the Java SDK and Maven are properly set up and the MongoDB server is up and running and listening to port `27017` for incoming connections.

How it works...

We will create test data for the recipes using the `setupUpdateTestData()` method. Here, we will simply put documents in the `javaTest` collection in the `javaDriverTest` database. We will add 20 documents in this collection with the value of `i` ranging from 1 to 20. This test data is used in different test case methods to create test data.

Let's now take a look at the methods in this class. We will first look at `basicUpdateTest()`. In this method, we will first create test data and then execute the following `update` method:

```
collection.update(
  new BasicDBObject("i", new BasicDBObject("$gt", 10)),
  new BasicDBObject("$set", new BasicDBObject("gtTen", true)));
```

The `update` method here takes two arguments: the first one is the query that will be used to select the eligible documents for the update, and the second parameter is the actual update. The first parameter looks confusing due to nested `BasicDBObject` instances; however, it is the `{'i' : {'$gt' : 10}}` condition, and the second parameter is the `{'$set' : {'gtTen' : true}}` update. The result of the update is an instance of `com.mongodb.WriteResult`. The instance of `WriteResult` tells us about the number of documents that got updated, the error that occurred while executing the write operation, and the write concern used for the update. Refer to the Java docs of the `WriteConcern` class for more details. This update, by default, only updates the first matching document, and only if multiple documents match the query.

The next method that we will look at is `multiUpdateTest`, which will update all the matching documents for the given query instead of the first matching document. The method we used on the collection instance is `updateMulti`. The `updateMulti` method is just a convenient method to update multiple documents. The following is the call that we will make to update multiple documents:

```
collection.updateMulti(new BasicDBObject("i",
    new BasicDBObject("$gt", 10)),
    new BasicDBObject("$set", new BasicDBObject("gtTen", true)));
```

The next operation that we will perform is to remove documents. The test case method to remove documents is `deleteTest()`. The documents are removed as follows:

```
collection.remove(new BasicDBObject(
    "i", new BasicDBObject("$gt", 10)),
    WriteConcern.JOURNALED);
```

We have two parameters here. The first one is the query for which matching documents will be removed from the collection. Note that all the matching documents will be removed by default, unlike in update, where only the first matching document will be removed by default. The second parameter is the write concern to be used for the remove operation.

Note that, when the server is started on a 32-bit machine, journaling is disabled by default. Using the journaling write concern on such machines causes the operation to fail with the following exception:

```
com.mongodb.CommandFailureException: { "serverUsed" :
"localhost/127.0.0.1:27017" , "connectionId" : 5 , "n" : 0 , "badGLE"
: { "getlasterror" : 1 , "j" : true} , "ok" : 0.0 , "errmsg" : "cannot
use 'j' option when a host does not have journaling enabled" , "code"
: 2}
```

This will need the server to be started with the `--journal` option. On 64-bit machines, this is not necessary, as journaling is enabled by default.

We will look at the `findAndModify` operation next. The test case method to perform this operation is `findAndModifyTest`. The following lines of code are used to perform this operation:

```
DBObject old = collection.findAndModify(
    new BasicDBObject("i", 10),
    new BasicDBObject("i", 100));
```

The operation is the query that will find the matching documents and then update them. The return type of the operation is an instance of `DBObject` before the update is applied. One important feature of the `findAndModify` operation is that the `find` and `update` operations are performed atomically.

Programming Language Drivers

The preceding method is a simple version of the `findAndModify` operation. There is an overloaded version of this method with the following signature:

```
DBObject findAndModify(DBObject query, DBObject fields, DBObject sort
,boolean remove, DBObject update, boolean returnNew, boolean upsert)
```

Let's see what these parameters are in the following table:

Parameter	Description
query	The `find` and `modify` operations have to find and modify the documents. This value of this parameter is the query that is used to query the documents that would be later modified.
fields	The `find` method supports a projection of the fields that it needs to be selected in the result document(s). The parameter here does the same job of selecting a fixed set of fields from the resulting document.
sort	If you haven't noticed already, let me tell you that the `sort` method can perform an atomic operation on only one document and also returns one document. This `sort` function can be used in cases where the query selects multiple documents, and only the first gets chosen for the operation. The `sort` operation is applied on the result before picking up the first document to update it.
remove	This is a Boolean flag that indicates whether to remove or update the document. If this value is `true`, the document will be removed.
update	This, unlike the `remove` attribute, is not a Boolean value but a `DBObject` instance that will tell what the update needs to be. Note that the removed Boolean flag gets precedence over this parameter. If the `remove` attribute is `true`, the update will not happen even if one is provided.
returnNew	The `find` operation returns a document, but which one? The one before the update was executed or the one after the update gets executed? When this Boolean flag is given as true, it returns the document after the update is executed.
upsert	This is a Boolean flag again that, when true, executes the `upsert` operation. It is relevant only when the intended operation is `update`.

There are more overloaded methods of this operation. Refer to the Java docs of `com.mongodb.DBCollection` for more methods. The `findAndModify` method we used ultimately invokes the method we discussed earlier with the fields and sort parameters as null and the remaining `remove`, `returnNew`, and `upsert` parameters being false.

Finally, we will look at query builder support in MongoDB Java API.

All the queries in Mongo are `DBObject` instances with possibly more nested `DBObject` instances in them. Things are simple for small queries, but they start getting ugly for more complicated queries. Consider a relatively simple query where we want to query for documents with i > 10 and i < 15. The Mongo query for this is {$and:[{i:{$gt:10}}, {i:{$lt:15}}] }. Writing this in Java using `BasicDBObject` instances is painful, and it looks as follows:

```
DBObject query = new BasicDBObject("$and",
  new BasicDBObject[] {
    new BasicDBObject("i", new BasicDBObject("$gt", 10)),
      new BasicDBObject("i", new BasicDBObject("$lt", 15))
  }
);
```

Thankfully, however, there is a class called `com.mongodb.QueryBuilder` that is a utility class to build complex queries. The preceding query is built using query builder as follows:

```
DBObject query = QueryBuilder.start("i").greaterThan(10).and("i").
lessThan(15).get();
```

This is less error-prone when writing a query and is easy to read as well. There are a lot of methods in the `com.mongodb.QueryBuilder` class, and I would like you to go through the Java docs of this class. The basic idea is to start construction using the start method and the key. We will then chain the method's calls to add different conditions and, when the addition of various conditions is done, the query is constructed using the `get()` method, which returns `DBObject`. Refer to the `queryBuilderSample` method in the test class for a sample usage of query builder support of MongoDB Java API.

See also

- *Chapter 5, Advanced Operations*, to know some more operations using GridFS and geospatial indexes and how to use them from the Java application with a small sample
- The Java docs for the current version of the MongoDB driver at `https://api.mongodb.org/java/current/`

Aggregation in Mongo using a Java client

The intention of this recipe is not to explain aggregation but to show how aggregation can be implemented using a Java client from a Java program. In this recipe, we will aggregate the data based on the state names and get the top five state names by the number of documents they appear in. We will make use of the `$project`, `$group`, `$sort`, and `$limit` operators for the process.

Getting ready

The test class used for this recipe is `com.packtpub.mongo.cookbook.MongoAggregationTest`. To execute the aggregation operations, we need to have a server up and running. A simple single node is what we will need. Refer to the *Single node installation of MongoDB* recipe in *Chapter 1, Installing and Starting the MongoDB Server*, to learn how to start the server. The data on which we will operate needs to be imported in the database. The steps to import the data are given in the *Creating test data* recipe in *Chapter 2, Command-line Operations and Indexes*. The next step is to download the `mongo-cookbook-javadriver` Java project from the book's website. Though Maven can be used to execute the test case, it is convenient to import the project in an IDE and execute the test case class. It is assumed that you are familiar with the Java programing language and comfortable using the IDE into which the project will be imported.

How to do it...

To execute the test case, one can either import the project in an IDE such as Eclipse and execute the test case or execute the test case from the command prompt using Maven.

If you are using an IDE, open the test class and execute it as a JUnit test case. If you plan to use Maven to execute this test case, go to the command prompt, change the directory to the root of the project, and execute the following command to execute this single test case:

```
$ mvn -Dtest=com.packtpub.mongo.cookbook.MongoAggregationTest test
```

Everything should execute fine if the Java SDK and Maven are properly set up and the MongoDB server is up and running and listening to port `27017` for incoming connections.

How it works...

The method used to look at aggregation functionality is `aggregationTest()` in our test class. The aggregation operation is performed on MongoDB from a Java client using the `aggregate()` method defined in the `DBCollection` class. The method has the following signature:

```
AggregationOutput aggregate(firstOp, additionalOps)
```

Only the first argument is mandatory; this forms the first operation in the pipeline. The second argument is a `varagrs` argument (a variable number of arguments with zero or more values) that allows more pipeline operators. All these arguments are of type `com.mongodb.DBObject`. If any exception occurs during the execution of the aggregation command, the aggregation operation will throw `com.mongodb.MongoException` with the cause of the exception.

The return type `com.mongodb.AggregationOutput` is used to get the result of the aggregation operation. From a developer's perspective, we are more interested in the results field of this instance, which can be accessed using the `results()` method of the returned object. The `results()` method returns an object of type `Iterable<DBObject>`, which one can iterate to get the results of the aggregation.

Let's look at how we implemented the aggregation pipeline in our test class:

```
AggregationOutput output = collection.aggregate(
  //{'$project':{'state':1, '_id':0}},
  new BasicDBObject("$project", new BasicDBObject("state",
1).append("_id", 0)),
  //{'$group':{'_id':'$state', 'count':{'$sum':1}}}
  new BasicDBObject("$group", new BasicDBObject("_id", "$state")
     .append("count", new BasicDBObject("$sum", 1))),
  //{'$sort':{'count':-1}}
  new BasicDBObject("$sort", new BasicDBObject("count", -1)),
  //{'$limit':5}
  new BasicDBObject("$limit", 5)
);
```

There are four operations in the pipeline in the following order. A `$project` operation, followed by `$group`, `$sort`, and then `$limit`.

The last two operations look inefficient; using them, we will sort everything but then just take the top five elements. The MongoDB server in such scenarios is intelligent enough to consider the limit operation while sorting; as a result of this, only the top five results need to be maintained rather than sorting all the results.

For Version 2.6 of MongoDB, the aggregation result can return a cursor. Though the preceding code snippet is still valid, the `AggregationResult` object is no longer the only way to get the results of the operation, but we can use `com.mongodb.Cursor` to iterate the results. Also, the preceding format is now deprecated in favor of the format that accepts a list of pipeline operators rather than `varargs` for the operators to be used. Refer to the Java docs of the `com.mongodb.DBCollection` class and look for various overloaded `aggregate()` methods.

MapReduce in Mongo using a Java client

In the previous recipe, we saw how to execute aggregation operations in Mongo using the Java client. In this recipe, we will work on the same use case as we did for the aggregation operation, but using MapReduce. The intent is to aggregate the data based on the state names and get the top five state names by the number of documents they appear in.

If you are not aware of how to write MapReduce code for Mongo from a programming language client and are seeing it for the first time, you might be surprised to see how it is actually done. You might have imagined that you will be writing the `map` and `reduce` functions in the programming language in which you are writing the code, Java in this case, and then using it to execute MapReduce. However, you need to bear in mind that MapReduce jobs run on the Mongo servers, and they execute JavaScript functions. Hence, irrespective of the programming language driver, the MapReduce functions are written in JavaScript. The programming language drivers just act as a means of letting us invoke and execute the MapReduce functions (written in JavaScript) on the server.

Getting ready

The test class used for this recipe is `com.packtpub.mongo.cookbook.MongoMapReduceTest`. To execute the MapReduce operations, we need to have a server up and running. A simple single node is what we will need. Refer to the *Single node installation of MongoDB* in *Chapter 1, Installing and Starting the MongoDB Server*, to learn how to start the server. The data on which we will operate needs to be imported in the database. The steps to import the data are given in the *Creating test data* recipe in *Chapter 2, Command-line Operations and Indexes*. The next step is to download the `mongo-cookbook-javadriver` Java project from the book's website. Though Maven can be used to execute the test case, it is convenient to import the project in an IDE and execute the test case class. It is assumed that you are familiar with the Java programming language and comfortable using the IDE to which the project will be imported.

How to do it...

To execute the test case, one can either import the project in an IDE such as Eclipse and execute the test case or execute the test case from the command prompt using Maven.

If you are using an IDE, open the test class and execute it as a JUnit test case. If you plan to use Maven to execute this test case, go to the command prompt, change the directory to the root of the project, and execute the following command to execute this single test case:

```
$ mvn -Dtest=com.packtpub.mongo.cookbook.MongoMapReduceTest test
```

Everything should execute fine if the Java SDK and Maven are properly set up and the MongoDB server is up and running and listening to port `27017` for incoming connections.

How it works...

The test case method for our MapReduce test is `mapReduceTest()`.

MapReduce operations can be done in Mongo from a Java client using the `mapReduce()` method defined in the `DBCollection` class. There are a lot of overloaded versions, and you might refer to Java docs of the `com.mongodb.DBCollection` class for more details on the various flavors of this method, but the one we used is as follows:

```
collection.mapReduce(mapper, reducer, output collection, query)
```

The method accepts four parameters:

- The first one is the mapper function, which is of type string and is a JavaScript code that will be executed on the Mongo database server
- The second one is the reducer function, which is of type string and is a JavaScript code that will be executed on the Mongo database server
- The third one is the name of the collection to which the output of the MapReduce execution will be written
- Finally, it is the query that will be executed by the server, and the result of this query will be the input to the MapReduce job execution

Since the assumption is that the reader is well versed in MapReduce operations from the shell, we won't explain the MapReduce JavaScript functions that we have in the test case method. However, it is pretty simple, and all it does is emit keys as the names of the states and values, which is the number of times the particular state name occurs. This resulting document with the key used, the state's name as the `_id` field, and another field called `value`, which is the sum of the times the particular state's name given in the `_id` field appeared in the collection, are added to the output collection, `javaMROutput` in this case. For example, in the entire collection, the state, Maharashtra, appeared 6446 times. Thus, the document for the state of Maharashtra is `{'_id': 'Maharashtra', 'value': 6446}`. To confirm whether this is the true value or not, you can execute the following query from the Mongo shell and see that the result is indeed 6446:

```
> db.postalCodes.count({state:'Maharashtra'})
6446
```

We are still not done as the requirement is to find the top five states by their occurrence in the collection. We still have just the states and their occurrences, so the final step is to sort the documents by the value field, which is the number of times the state's name occurred in descending order, and limit the result to five documents.

See also

Chapter 8, Integration with Hadoop, for different recipes on executing MapReduce jobs on MongoDB using the Hadoop connector. This allows us to write the `map` and `reduce` functions in languages such as Java and Python.

4
Administration

In this chapter, we will cover the following recipes related to MongoDB administration:

- Renaming a collection
- Viewing collection stats
- Viewing database stats
- Disabling the preallocation of data files
- Manually padding a document
- Understanding the mongostat and mongotop utilities
- Estimating the working set
- Viewing and killing the currently executing operations
- Using profiler to profile operations
- Setting up users in MongoDB
- Understanding interprocess security in MongoDB
- Modifying collection behavior using the collMod command
- Setting up MongoDB as a Windows Service
- Configuring a replica set
- Stepping down as a primary instance from the replica set
- Exploring the local database of a replica set
- Understanding and analyzing oplogs
- Building tagged replica sets
- Configuring the default shard for nonsharded collections
- Manually splitting and migrating chunks
- Performing domain-driven sharding using tags
- Exploring the config database in a sharded setup

Administration

Renaming a collection

Have you ever come across a scenario where you have named a table in a relational database, and at a later point of time, felt that the name could have been better? Or perhaps, the organization you work for was late in realizing that the table names are really getting messy and wants to enforce some standards on the names? Relational databases do have some proprietary ways to rename the tables, and a database admin can do that for you.

This raises a question though. In the Mongo world, where collections are synonymous with tables, is there a way to rename a collection after it is created? In this recipe, we will explore this feature of Mongo, where we rename an existing collection with some data in it.

Getting ready

Running a MongoDB instance is what we will need for performing this collection renaming experiment. Refer to the *Single node installation of MongoDB* recipe in *Chapter 1, Installing and Starting the MongoDB Server*, for how to start the server. The operations we will be performing would be from the Mongo shell.

How to do it...

Let's take a look at the steps in detail:

1. Once the server is started, and assuming it is listening for client connections on the default port 27017, execute the following command to connect to it from the shell:

   ```
   > mongo
   ```

2. Once connected, using the default test database, let us create a collection with some test data. The collection we will be using is named sloppyNamedCollection:

   ```
   > for(i = 0 ; i < 10 ; i++) db.sloppyNamedCollection.insert({'i':i})
   ```

3. The test data will now be created (we may verify the data by querying the sloppyNamedCollection collection).

4. Rename the collection as neatNamedCollection using the following command:

   ```
   > db.sloppyNamedCollection.renameCollection('neatNamedCollection')
   { "ok" : 1 }
   ```

5. Verify that the sloppyNamedCollection collection is no longer present, by executing the following command:

   ```
   > show collections
   ```

6. Finally, query the neatNamedCollection collection to verify that the data that was originally in sloppyNamedCollection is indeed present in it. Simply execute the following command on the Mongo shell:

```
> db.neatNamedCollection.find()
```

How it works...

Renaming a collection is pretty simple. It is accomplished with the renameCollection method, which takes two arguments. Generally, the function signature is as follows:

```
> db.<collection to rename>.renameCollection('<target name of the collection>', <drop target if exists>)
```

The first argument is the name by which the collection is to be renamed.

The second parameter that we didn't use is a Boolean value that tells the command whether to drop the target collection (if it exists) or not. This value defaults to false, which means the target must not be dropped but must give an error instead. This is a sensible default, else the results would be ghastly if we accidentally gave a collection name that exists and didn't wish to drop it. If however, you know what you are doing and want the target to be dropped while renaming the collection, pass the second parameter as true. The name of this parameter is dropTarget. In our case, the call would have been:

```
> db.sloppyNamedCollection.renameCollection('neatNamedCollection', true)
```

Now, as an exercise, try creating sloppyNamedCollection again and rename it without the second parameter (or false as the value). You should see Mongo complaining that the target namespace exists. Then, rename it again with the second parameter as true. This time, the renaming operation executes successfully.

Note that the rename operation will keep the original and the newly renamed collection in the same database. This renameCollection method is not enough to move/rename the collection across another database. In such cases, we need to run the renameCollection command as follows:

```
> db.runCommand({ renameCollection: "<source_namespace>", to: "<target_namespace>", dropTarget: <true|false> });
```

In our case, suppose we want to move sloppyNamedCollection from the test database to newDatabase, rename it neatNamedCollection, and drop the target database if it exists; we will execute the following command:

```
> db.runCommand({ renameCollection: "test.sloppyNamedCollection", to: "newDatabase.neatNamedCollection", dropTarget: true });
```

Also, note that the rename collection operation doesn't work on sharded collections.

Administration

Viewing collection stats

When it comes to the usage of storage, one of the interesting statistics from an administrative point of view is perhaps the number of documents in a collection, possibly to estimate future space and memory requirements based on the growth of the data to get high-level statistics of the collection.

Getting ready

To find the stats of the collection, we need to have a server up and running, and a single node should be ok. Refer to the *Single node installation of MongoDB* recipe in *Chapter 1, Installing and Starting the MongoDB Server*, for how to start the server. The data on which we will be operating needs to be imported into the database. The steps to import the data are given in the *Creating test data* recipe in *Chapter 2, Command-line Operations and Indexes*. Once these steps are completed, we are all set to go ahead with this recipe.

How to do it...

We will be using the `postalCodes` collection to view the stats. Let's take a look at the steps in detail:

1. Open the Mongo shell and connect it to the running MongoDB instance. In this case, start Mongo on the default port `27017` and execute the following command:

    ```
    $ mongo
    ```

2. With the data imported, create an index in the `pincode` field, if one doesn't exist, as follows:

    ```
    > db.postalCodes.ensureIndex({'pincode':1})
    ```

3. On the Mongo terminal, execute the following command:

    ```
    > db.postalCodes.stats()
    ```

4. Observe the output. Now execute the following command on the shell:

    ```
    > db.postalCodes.stats(1024)
    {
      "ns" : "test.postalCodes",
      "count" : 39732,
      "size" : 5561,
      "avgObjSize" : 0.1399627504278667,
      "storageSize" : 16380,
      "numExtents" : 6,
    ```

Chapter 4

```
            "nindexes" : 2,
            "lastExtentSize" : 12288,
            "paddingFactor" : 1,
            "systemFlags" : 1,
            "userFlags" : 0,
            "totalIndexSize" : 2243,
            "indexSizes" : {
               "_id_" : 1261,
               "pincode_1" : 982
            },
            "ok" : 1
   }
```

Again, observe the output. We will now see what these values mean to us in the next section.

How it works...

If we observe the output for the `db.postalCodes.stats()` and `db.postalCodes.stats(1024)` commands, we see that the second one has all the figures in KB whereas the first one is in bytes. The parameter provided is known as scale and all the figures indicating size are divided by this scale. In this case, as we gave the value as `1024`, we get all the values in KB; whereas if 1024 * 1024 is passed as the value of the scale, the size shown will be in MB. For our analysis, we will use the one that shows the sizes in KB:

```
> db.postalCodes.stats(1024)
{
   "ns" : "test.postalCodes",
   "count" : 39732,
   "size" : 5561,
   "avgObjSize" : 0.1399627504278667,
   "storageSize" : 16380,
   "numExtents" : 6,
   "nindexes" : 2,
   "lastExtentSize" : 12288,
   "paddingFactor" : 1,
   "systemFlags" : 1,
   "userFlags" : 0,
   "totalIndexSize" : 2243,
   "indexSizes" : {
```

Administration

```
    "_id_" : 1261,
    "pincode_1" : 982
  },
  "ok" : 1
}
```

The following table shows the meaning of the important fields:

Field	Description
ns	This is the fully qualified name of the collection with the `<database>.<collection name>` format.
count	This is the number of documents in the collection.
size	This is the actual storage size occupied by the documents in the collection. Adding, deleting, or updating documents in the collection can change this figure. The scale parameter affects this field's value and in our case, this value is in KB as `1024` is the scale.
avgObjSize	This is the average size of the document in the collection. It is simply the `size` field divided by the count of documents in the collection. The scale parameter affects this field's value and in our case, this value is in KB as `1024` is the scale.
storageSize	Mongo preallocates the space on the disk to ensure that the documents in the collection are kept on continuous locations to provide better performance in disk access. This preallocation fills up the files with zeros and then starts allocating space to these inserted documents. This field reveals the size of the storage used by this collection. This figure will generally be much more than the actual size of the collection. The scale parameter affects this field's value and in our case, this value is in KB as `1024` is the scale.
numExtents	As we saw, Mongo preallocates continuous disk space to the collections for performance purposes. However, as the collection grows, new space needs to be allocated. This field gives the number of such continuous chunk allocation. This continuous chunk is called extent.
nindexes	This field gives the number of indexes present in the collection. This value would be 1, even if we do not create an index on the collection, as Mongo implicitly creates an index on the `_id` field.
lastExtentSize	This is the size of the last extent allocated. The scale parameter affects this field's value and in our case, this value is in KB as `1024` is the scale.

Field	Description
`paddingFactor`	We can look at this factor as a multiplier to the actual document size in order to compute the storage size. For example, if the document to be inserted is 2 KB, with a `paddingFactor` field of 1, the size allocated to the document is 2 KB; that is, with no padding. On the other hand, if the `paddingFactor` field is `1.5`, the space allocated to the document will be 3 KB (2 * 1.5), which gives a padding of 1 KB. In our case, the `paddingFactor` field is `1` because we did a `mongoimport`. We will discuss padding and padding factor in the next section.
`totalIndexSize`	Indexes take up space to store. This field gives the total size taken up by the indexes on the disk. The scale parameter affects this field's value and in our case, this value is in KB as `1024` is the scale.
`indexSizes`	This field itself is a document, with the key as the name of the index and the value as the size of the index in question. In our case, we had created an index explicitly on the `pincode` field. Thus, we see the name of the index as the key and the size of the index on disk as the value. The total of these values of all the indexes is the same as the value given earlier, that is, `totalIndexSize`. The scale parameter affects this field's value and in our case, this value is in KB as `1024` is the scale.

Let's take a look at the `paddingFactor` field. Documents are placed on the storage device in continuous locations. If, however, an update occurs that causes the size of the document to increase, Mongo obviously will not be able to increase the document size if there was no buffer space kept after the document. The only solution is to copy the entire document towards the end of the collection with the necessary updates made to it. This operation turns out to be expensive, affecting the performance of such update operations. If the `paddingFactor` field is `1`, no padding or buffer space is kept between two consecutive documents, making it impossible for the first of these two documents to grow on updates. If this `paddingFactor` field is more than 1, there would be some buffer space accommodating some small size changes for the documents. This `paddingFactor` field, however, is not set by the user and MongoDB calculates it for the collection over a period of time. It then uses this calculated `paddingFactor` field to allocate a space for the new documents inserted. To get a feel of how this padding factor changes, let us do a small exercise:

1. Execute the following command in the Mongo shell:

   ```
   > for(i = 0 ; i < 10; i++) {
         db.paddingFactorTest.insert({value:'Hello World'})
     }
   ```

2. Now execute the following command and take note of the `paddingFactor` value (it would be 1):

   ```
   > db.paddingFactorTest.stats()
   ```

3. We will now make some updates to let the document grow in size as follows:

```
> for(i = 0; i < 5;i++) {
        db.paddingFactorTest.update({value:'Hello World'},
   {$push:{value1:'Value'}}, false, true)
    }
> db.paddingFactorTest.stats()
```

Query the stats again and observe the value of `paddingFactor` that has gone slightly over 1, which shows that the MongoDB server adjusted this value while allocating space for a document insertion at a later point in time.

We saw how `paddingFactor` affects the storage allocated to a document, but neither do we have control on this value, nor can we instruct Mongo beforehand on what additional buffer needs to be allocated to each document inserted based on the anticipated growth of a document. There is, however, a technique that let us achieves this in a way that we will see in the *Manually padding a document* recipe.

See also

- The *Viewing database stats* recipe to view the stats at a database level

Viewing database stats

In the previous recipe, we saw how to view some important statistics of a collection from an administrative perspective. In this recipe, we'll get an even clearer picture; getting those (or most of those) statistics at the database level.

Getting ready

To find the stats of the database, we need to have a server up and running, and a single node should be ok. Refer to the *Single node installation of MongoDB* recipe in *Chapter 1, Installing and Starting the MongoDB Server*, for how to start the server. The data on which we would be operating needs to be imported into the database. The steps to import the data are given in the *Creating test data* recipe in *Chapter 2, Command-line Operations and Indexes*. Once these steps are completed, we are all set to go ahead with this recipe. Refer to the previous recipe, *Viewing collection stats*, if you need to see how to view stats at the collection level.

How to do it...

We will be using the `test` database for the purpose of this recipe. It already has the `postalCodes` collection in it. Let's take a look at the steps in detail:

1. Connect to the server using the Mongo shell by typing in the following command from the operating system terminal (it is assumed that the server is listening to port `27017`):

   ```
   $ mongo
   ```

2. On the shell, execute the following command and observe the output:

   ```
   > db.stats()
   ```

3. Now, execute the following command, but this time with the scale parameter (observe the output):

   ```
   > db.stats(1024)
   {
       "db" : "test",
       "collections" : 3,
       "objects" : 39738,
       "avgObjSize" : 143.32699179626553,
       "dataSize" : 5562,
       "storageSize" : 16388,
       "numExtents" : 8,
       "indexes" : 2,
       "indexSize" : 2243,
       "fileSize" : 196608,
       "nsSizeMB" : 16,
       "dataFileVersion" : {
           "major" : 4,
           "minor" : 5
       },
       "ok" : 1
   }
   ```

Administration

How it works...

Let us start by looking at the collections field. If you look carefully at the number and also execute the show collections command on the Mongo shell, you will find one extra collection in the stats as compared to those achieved by executing the command. The difference is denotes one collection, which is hidden, and its name is system.namespaces. You may execute db.system.namespaces.find() to view its contents.

Getting back to the output of stats operation on the database, the objects field in the result has an interesting value too. If we find the count of documents in the postalCodes collection, we see that it is 39732. The count shown here is 39738, which means there are six more documents. These six documents come from the system.namespaces and system.indexes collection. Executing a count query on these two collections will confirm it. Note that the test database doesn't contain any other collection apart from postalCodes. The figures will change if the database contains more collections with documents in it.

The scale parameter, which is a parameter to the stats function, divides the number of bytes with the given scale value. In this case, it is 1024, and hence, all the values will be in KB. Let's analyze the output:

```
> db.stats(1024)
{
  "db" : "test",
  "collections" : 3,
  "objects" : 39738,
  "avgObjSize" : 143.32699179626553,
  "dataSize" : 5562,
  "storageSize" : 16388,
  "numExtents" : 8,
  "indexes" : 2,
  "indexSize" : 2243,
  "fileSize" : 196608,
  "nsSizeMB" : 16,
  "dataFileVersion" : {
    "major" : 4,
    "minor" : 5
  },
  "ok" : 1
}
```

The following table shows the meaning of the important fields:

Field	Description
db	This is the name of the database whose stats are being viewed.
collections	This is the total number of collections in the database.
objects	This is the count of documents across all collections in the database. If we find the stats of a collection by executing db.<collection>.stats(), we get the count of documents in the collection. This attribute is the sum of counts of all the collections in the database.
avgObjectSize	This is simply the size (in bytes) of all the objects in all the collections in the database, divided by the count of the documents across all the collections. This value is not affected by the scale provided even though this is a size field.
dataSize	This is the total size of the data held across all the collections in the database. This value is affected by the scale provided.
storageSize	This is the total amount of storage allocated to collections in this database for storing documents. This value is affected by the scale provided.
numExtents	This is the count of all the extents in the database across all the collections. This is basically the sum of numExtents in the collection stats for collections in this database.
indexes	This is the sum of indexes across all collections in the database.
indexSize	This is the size (in bytes) for all the indexes of all the collections in the database. This value is affected by the scale provided.
fileSize	This is simply the addition of the size of all the database files you should find on the filesystem for this database. The files will be named test.0, test.1, and so on for the test database. This value is affected by the scale provided.
nsSizeMB	This is the size of the file in MBs for the .ns file of the database.

Another thing to note is the value of avgObjectSize, and there is something weird in this value. Unlike this very field in the collection's stats, which is affected by the value of the scale provided, in database stats this value is always in bytes, which is pretty confusing and one cannot really be sure why this is not scaled according to the provided scale.

See also

- *Instant MongoDB, Packt Publishing* (https://www.packtpub.com/big-data-and-business-intelligence/instant-mongodb-instant)

Disabling the preallocation of data files

Data files are preallocated in Mongo and filled with zeros even before data is inserted into collections to prevent disk fragmentation. These data files are allocated starting from 64 MB for the first, 128 MB for the second, 256 MB for the third, and so on, till a maximum size of 2 GB after which all files would be 2 GB. Though these preallocated data files prevent disk fragmentation, prepopulating requires time and consumes disk space. However, preallocating such files just when the data is inserted can take a significant amount of time and thus, Mongo preallocates an additional file in the background and keeps an additional datafile ready. If, however, this preallocation is not desired for, say, a `test` database, where quick startup is desired and less disk space consumption is more important, this preallocation can be disabled. This, however, should not be done on production systems.

How to do it...

When starting a server, we can start the MongoDB server with the `--noprealloc` flag to disable this preallocation. For instance, a server started to listen on the default port with preallocation disabled will be started as follows:

```
$ mongod --noprealloc
```

Manually padding a document

Without getting too much into the internals of storage, MongoDB uses memory-mapped files, which means that the data is stored in files exactly as it would be in memory; it will use low-level OS services to map these pages to memory. The documents are stored in continuous locations in Mongo data files and the problem arises when the document grows and no longer fits in the space. In such scenarios, Mongo rewrites the document towards the end of the collection with the updated data and clears up the space where it was originally placed (note that this space is not released to the OS as free space).

This is not a big problem for applications, which don't expect the documents to grow in size; however, this is a big performance hit for those who foresee this growth in the document size over a period of time and potentially, a lot of such document movements. The `paddingFactor` field, that we saw in the *Viewing collection stats* recipe, gets updated over a period of time, to some extent, and allocates some buffer for the document to grow. However, this is only over a period of time once a lot of documents have already been moved across the collection and the MongoDB server adjusts the padding size. Moreover, at the time of writing, this padding factor cannot be set in any way for the collection beforehand, based on your anticipated increase in the size of the document, to counter this document's rewrites by Mongo, and is set to a default value of 1. However, there is a small trick that does let you do this, and that is what we will see in this recipe. This is a commonly used practice for such requirements.

Getting ready

Nothing is particularly needed for this recipe, unless you plan to try out this simple technique; in which case, you would need a single instance up and running. Refer to the *Single node installation of MongoDB* recipe in *Chapter 1, Installing and Starting the MongoDB Server*, for how to start the server.

How to do it...

The idea of this technique is to add some dummy data to the document that is to be inserted. This dummy data's size, in addition to other data in the document, is approximately the same as the anticipated size of the document.

For example, if the average size of the document is estimated to be around 1200 bytes over a period of time, and there is 300 bytes of data present in the document while inserting it, we will add a dummy field that is around 900 bytes in size, so that the total document size sums up to 1200 bytes.

Once the document is inserted, we unset this dummy field, which leaves a hole in the file between the two consecutive documents. This empty space will then be used when the document grows over a period of time, minimizing the document's movements. This is not a foolproof method, as any document growing beyond the anticipated average growth will have to be copied by the server to the end of the collection. Also, documents not growing to the anticipated size tend to waste disk space.

The applications can come up with an intelligent strategy to, perhaps, adjust the size of the padding field based on a field in the document to take care of these shortcomings. However, this is something that is up to the application developers.

Let us now see a sample of this approach:

1. We define a small function that will add a field called `padField` with an array of string values to the document as follows:

   ```
   function padDocument(doc) {
     doc.padField = []
     for(i = 0 ; i < 20 ; i++) {
       doc.padField[i] = 'Dummy'
     }
   }
   ```

   It will add an array called `padField` and a string called `Dummy` 20 times. There is no restriction on what type you add to the document and how many times it is added as long as it consumes the space you desire. The preceding code snippet is just a sample.

Administration

2. The next step is to insert a document. We will define another function called `insert` in the following manner:

```
function insert(collection, doc) {
  //1. Pad the document with padField
  padDocument(doc);
  //2. Create or store the _id field that would be used later
  if(typeof(doc._id) == 'undefined') {
    _id = ObjectId()
    doc._id = _id
  }
  else {
    _id = doc._id
  }
  //3. Insert the document with the padded field
  collection.insert(doc)
  //4. Remove the padded field. Use the saved _id to find the document to be updated.
  collection.update({'_id':_id}, {$unset:{'padField':1}})
}
```

3. We will now put this in to action by inserting a document in the `testCol` collection in the following manner:

   ```
   insert(db.testCol, {i:1})
   ```

4. You may query the `testCol` collection using the following query and check whether the inserted document exists or not:

   ```
   > db.testCol.findOne({i:1})
   ```

Note that on querying, you would not find `padField` in the `testCol` collection. However, the space once occupied by the array stays between the subsequently inserted documents even if the field was unset.

How it works...

The `insert` function is self-explanatory and has comments in it to tell you what it does. An obvious question is, how can we be sure this is indeed what we intended to do? For this purpose, we shall do a small activity as follows. We will work on a `manualPadTest` collection for this purpose. From the Mongo shell, execute the following commands:

```
> db.manualPadTest.drop()
> db.manualPadTest.insert({i:1})
```

Chapter 4

```
> db.manualPadTest.insert({i:2})
> db.manualPadTest.stats()
```

Take note of the `avgObjSize` field in the stats. Next, execute the following commands from the Mongo shell:

```
> db.manualPadTest.drop()
> insert(db.manualPadTest , {i:1})
> insert(db.manualPadTest , {i:2})
> db.manualPadTest.stats()
```

Take note of the `avgObjSize` field in the stats. This figure is much larger than the one we saw earlier in a regular insert without padding. The `paddingFactor` field, as we see in both cases, still is 1, but the latter case has more buffer for the document to grow.

One catch in the `insert` function we used in this recipe is that the insert into the collection and the update document operations are not atomic.

Understanding the mongostat and mongotop utilities

Most of you might find these names similar to two popular Unix commands, `iostat` and `top`. For MongoDB, `mongostat` and `mongotop` are two utilities that do pretty much the same job as the two Unix commands, and there is no prize for guessing that these are used to monitor the Mongo instance.

Getting ready

In this recipe, we will be simulating some operations on a standalone Mongo instance by running a script that will attempt to keep your server busy; then, in another terminal, we will be running these utilities to monitor the `db` instance.

You need to start a standalone server listening to any port for client connections; in this case, we will stick to the default `27017`. In case you are not aware of how to start a standalone server, refer to the *Single node installation of MongoDB* recipe in *Chapter 1, Installing and Starting the MongoDB Server*. We also need to download the `KeepServerBusy.js` script from the book's website and keep it handy for execution on the local drive. Also, it is assumed that the `bin` directory of your Mongo installation is present in the `path` variable of your operating system. If not, then these commands need to be executed with the absolute path of the executable from the shell. The `mongostat` and `mongotop` utilities come as standard with the Mongo installation.

Administration

How to do it...

Let's take a look at the steps in detail:

1. Start the MongoDB server. Let it listen to the default port for connections.
2. In a separate terminal, execute the `KeepServerBusy.js` JavaScript as follows:

   `$ mongo KeepServerBusy.js --quiet`

3. Open a new OS terminal and execute the following command:

   `$ mongostat`

4. Capture the output content for some time and then hit *Ctrl* + *C* to stop the command from capturing more stats. Keep the terminal open or copy the stats to another file.
5. Now execute the following command from the terminal:

   `$ mongotop`

6. Capture the output content for some time and then hit *Ctrl* + *C* to stop the command from capturing more stats. Keep the terminal open or copy the stats to another file.
7. Hit *Ctrl* + *C* in the shell, where the `KeepServerBusy.js` JavaScript was executed, to stop the operation that keeps the server busy.

How it works...

Let us see what we have captured from these two utilities. We start by analyzing `mongostat`. On my laptop, the output of the `$ mongostat` command is as follows:

```
$ mongostat
connected to: 127.0.0.1
insert   query  update  delete  getmore  command  flushes  mapped    vsize    res
faults     locked db
idx miss %     qr|qw    ar|aw   netIn   netOut         conn     time
1000     1     1000     1794    1             1|0              0     320m
808m     54m   37       test:85.7%             0              0|0         0|1
271k           94k      2       23:24:30

2000     1     1326     1206    1             1|0              0     320m
808m     54m   113      test:83.3%             0              0|0         0|1
339k           51k      2       23:24:31

1000     1      952     1000    1             1|0              0     320m    808m
54m      28    test:84.4%             0              0|0      0|1       219k       51k
2        23:24:32

77       1      722     1000    1             1|0              0     320m    808m
54m      87    test:73.0%             0              0|0      0|1       131k       51k
2        23:24:33
```

Chapter 4

```
923          1      1000    792            1         1|0              0           320m
808m        54m     42    test:83.3%              0              0|0        0|0         206k
51k          2     23:24:34
1000         1      1000    934            1         1|0              0           320m    808m
54m         150    test:84.6%        0                     0|0        0|1        220k       51k
2           23:24:35
1000         1      1000    920            1         1|0              0           320m    808m
54m          13    test:84.9%        0                     0|0        0|1        219k       51k
2           23:24:36
```

You may choose to look at what the `KeepServerBusy.js` script is doing to keep the server busy. All it does is insert 1000 documents in the `monitoringTest` collection; update them one by one to set a new key in them; execute a find and iterate through all of them; and finally, delete them one by one. Basically, it is a write-intensive operation.

The output does look ugly with the content wrapping, but let us analyze the fields one by one and see what to look out for. The following table gives a description of each column:

Column(s)	Description
`insert`, `query`, `update`, and `delete`	These are the first four columns indicating the number of insert, query, update, and delete operations per second. It is per second as the time frame in which these figures are captured is separated by 1 second, which is indicated by the last column.
`getmore`	This is used when the cursor runs out of data for the query; it executes a `getmore` operation on the server to get more results for the query executed earlier. This column shows the number of `getmore` operations executed in this given time frame of 1 second. In our case, not many `getmore` operations are executed.
`command`	This shows the number of commands executed on the server in the given time frame of 1 second. In our case, it wasn't much and was only 1. The number after a | is 0 in our case as this was in the standalone mode. Try executing `mongostat` connecting to a replica set primary and secondary. You should see slightly different figures there.
`flushes`	This is the number of times data was flushed to the disk in an interval of 1 second.
`mapped`, `vsize`, and `res`	Mapped memory is the amount of memory mapped by the Mongo process to the database. This typically will be same as the size of the database. Virtual memory on the other hand is the memory allocated to the entire `mongod` process. This typically will be more than twice the size of mapped memory, especially when journaling is enabled. The resident memory is the physical memory used by Mongo. All these figures are given in MB. The total amount of physical memory might be a lot more than what is being used by Mongo, but that is not a concern unless a lot of page faults occur, (we saw this in the preceding point).

Column(s)	Description
`faults`	These are the number of page faults occurring per second. These numbers should be as low as possible. It indicates the number of times Mongo had to go to disk to obtain the document/index that was missing in the main memory. This problem is not as big a problem when using SSD for persistent storage as it is when using spinning disk drives.
`locked`	From version 2.2, all write operations to a collection lock the database in which the collection is and do not acquire a global-level lock. This field shows the database that was locked for the majority of time in a given time interval. In our case, the `test` database is locked.
`idx miss %`	This field gives the number of times a particular index was needed and was not present in memory. This causes a page fault, and the disk needs to be accessed to get the index. Another disk access might be needed to get the document as well. This figure too should be low. A high percentage of index misses is something that will need attention.
`qr|qw`	These are the queued-up reads and writes that are waiting for the chance to be executed. If this number goes up, it shows that the database is getting overwhelmed by the volume of reads and writes, which are more than it can handle. The page faults with memory stats and a database lock percentage are some of the stats that need to be examined as well if this figure is high. If the dataset is too large, sharding the collection can improve the performance significantly.
`ar|aw`	This is the number of active readers and writers (clients). Not something to worry about even for a large number, as long as other stats we saw earlier are under control.
`netIn` and `netOut`	This is the network traffic in and out of the MongoDB server in the given time frame. The figure is in bits. For example, 271 kbit means 271 kilobits.
`conn`	This indicates the number of open connections. Something to keep a watch on to see this doesn't keep getting higher.
`time`	This is the time interval when this sample was captured.

There are some more fields seen if `mongostat` is connected to a replica set primary or secondary. As an assignment, once the stats or a standalone instance are collected, start a replica set server and execute the same script to keep the server busy. Use `mongostat` to connect to primary and secondary instances and see if different stats are captured.

Chapter 4

Apart from `mongostat`, we also used the `mongotop` utility to capture the stats. Let us see its output and make some sense out of it:

```
$ mongotop
connected to: 127.0.0.1
```

ns	total	read	write
2014-01-15T17:55:13			
test.monitoringTest	899ms	1ms	898ms
test.system.users	0ms	0ms	0ms
test.system.namespaces	0ms	0ms	0ms
test.system.js	0ms	0ms	0ms
test.system.indexes	0ms	0ms	0ms

ns	total	read	write
2014-01-15T17:55:14			
test.monitoringTest	959ms	0ms	959ms
test.system.users	0ms	0ms	0ms
test.system.namespaces	0ms	0ms	0ms
test.system.js	0ms	0ms	0ms
test.system.indexes	0ms	0ms	0ms

ns	total	read	write
2014-01-15T17:55:15			
test.monitoringTest	954ms	1ms	953ms
test.system.users	0ms	0ms	0ms
test.system.namespaces	0ms	0ms	0ms
test.system.js	0ms	0ms	0ms
test.system.indexes	0ms	0ms	0ms

Administration

There is not much to look at in this stat. We see the total time for which the database was busy reading or writing in the given slice of 1 second. The value given in the total will be the sum of the read and the write time. If we actually compare the `mongotop` and `mongostat` utilities for the same time slice, the percentage of time duration for which the write was taking place will be very close to the figure given in the percentage time the database was locked in the mongostat's output.

The `mongotop` command accepts a parameter on the command line as follows:

```
$ mongotop 5
```

In this case, the interval after which the stats will be printed out will be 5 seconds, as against the default value of 1 second.

See also

- The *Estimating the working set* recipe, to learn how to estimate the working set using the working set estimator command introduced in Mongo 2.4
- The *Viewing and killing the currently executing operations* recipe, to learn how to get the current executing operations from the shell and kill them if needed
- The *Using profiler to profile operations* recipe to learn how to use the in-built profiling feature of Mongo to log operation's execution time

Estimating the working set

We start by defining what the working set is. It is a subset of the total data frequently accessed by the application. In an application, which stores information over a period of time, the working set is mostly the recently accessed data. The word "recently" is subjective; for some it might be a day or two, for others it might be a couple of months. This is mostly something that needs to be thought of while designing the application and sizing the database. The working set is something that needs to be in the RAM of the database server to minimize the page faults and get the optimum performance.

In this recipe, we will see a way that gives the estimate of your working set and is a feature introduced in Mongo 2.4. The word "estimator" is slightly misleading, as the initial sizing still is a manual activity, and the system designers need to be judicious about the server configuration. The working set estimator utility we will see now is more of a reactive approach, which will kick in once the application is up and running. It provides metrics that can be used by monitoring tools, and tells us if the RAM on the server can accommodate the working set or if the set outruns the available RAM. This then demands some resizing of the hardware or scaling of the database horizontally.

Getting ready

In this recipe, we will be simulating some operations on a standalone Mongo instance. We need to start a standalone server listening to any port for client connections; in this case, we will stick to the default `27017`. In case you are not aware of how to start a standalone server, refer to the *Single node installation of MongoDB* recipe in *Chapter 1, Installing and Starting the MongoDB Server*. Connect to the server from the Mongo shell.

How to do it...

The working set now is a part of the server's status output. There is a field called `workingSet`, whose value is a document that gives these estimates.

This working set is not available as part of the standard `serverStatus` command and needs to be demanded explicitly. It is not an operation cheap on resources, and thus needs to be monitored if it is executed frequently. Frequent invocations can have a detrimental effect on the performance of the server.

We need to run the following command from the Mongo shell to get the working set estimates:

```
> db.runCommand({serverStatus:1, workingSet:1}).workingSet
{
        "note" : "thisIsAnEstimate",
        "pagesInMemory" : 6188,
        "computationTimeMicros" : 11524,
        "overSeconds" : 3977
}
```

How it works...

There are just four fields in this document for the working set estimate, with the first just stating in text that this is an estimate. The `pagesInMemory`, `computationTimeMicros`, and `overSeconds` fields are something we will be more interested in.

Administration

We will look at the `overSeconds` field first. This is the time in seconds between the first and the last page loaded by Mongo in the memory. When the server is started, this value will obviously be less but eventually, with more data being accessed with time, more pages will be loaded by Mongo in the memory. If the RAM available is abundant, the first loaded page will stay in memory and new pages will continue to load as and when needed. Hence, the time will also increase, as the difference between the most recently loaded page and the oldest page will increase. If this time stays low, or even decreases, we can say that the oldest and newest page in Mongo were loaded in just the number of seconds given by this figure. This can be an indication that the number of pages accessed and loaded in memory by the MongoDB server is more than those that can be held in memory. As Mongo uses the **least recently used** (**LRU**) policy to evict a page from the memory to make space for the new page, we possibly are risking evicting pages that might be needed again, causing more page faults.

This is where the `pagesInMemory` field comes in. This tells us, over a period of time, the number of pages Mongo loaded in the memory. Each page multiplied by around 4 KB gives the size of data loaded in the memory in bytes. Thus, if all data is being accessed after the server is started, this size will be around your data size. This number will keep increasing with time hence this field, in conjunction with the `overSeconds` field, is an important statistic.

The final field, `computationTimeMicros`, gives the time in microseconds taken by the server to give this statistic for the working set. As we can see, it is not an incredibly cheap operation to execute and thus, this statistic should be demanded with caution, especially on high-throughput systems.

Viewing and killing the currently executing operations

In this recipe, we will see how to view the current running operations and kill some operations that have been running for a long time.

Getting ready

In this recipe, we will simulate some operations on a standalone Mongo instance. We need to start a standalone server listening to any port for client connections; in this case, we will stick to the default `27017`. If you are not aware of how to start a standalone server, refer to the *Single node installation of MongoDB* recipe in *Chapter 1, Installing and Starting the MongoDB Server*. We also need to start two shells connected to the server started. One shell will be used for background index creation, and the other will be used to monitor the current operation and then kill it.

How to do it...

Unlike in our test environment, we will not be able to simulate the actual long-running operation. We, however, will try to create an index and hope it takes a long time to create. Depending on your target hardware configuration, the operation may take some time. Let's see the steps in detail:

1. To start this test, let us execute the following command on the Mongo shell:

    ```
    > db.currentOpTest.drop()

    > for(i = 1 ; i < 10000000 ; i++) { db.currentOpTest.insert({'i':i}) }
    ```

    The preceding insertion might take some time to insert 10 million documents.

    Once the documents are inserted, we will execute an operation that will create the index in the background. If you would like to know more about index creation, refer to the recipe *Background and foreground index creation from the shell* in *Chapter 2, Command-line Operations and Indexes*, but it is not a prerequisite for this recipe.

2. Create a background index on the `i` field in the document. This index-creation operation is what we will be viewing from the `currentOp` operation and is what we will attempt to kill by using the kill operation. Execute the following command in one shell to initiate the background index creation operation:

    ```
    > db.currentOpTest.ensureIndex({i:1}, {background:1})
    ```

    This takes a fairly long time; on my laptop, it took well over 100 seconds

3. In the second shell, execute the following command to get the current executing operations:

    ```
    > db.currentOp().inprog
    ```

4. Take a note of the in-progress operations and find the one for index creation. In our case, on the test machine, it was the only one in progress. It will be an operation on `system.indexes` and the operation will be `insert`. The keys to look out for in the output document are `ns` and `op` respectively. We need to note the first field, namely `opid`, of this operation. In this case, it is `11587458`. The sample output of the command is given in the next section.

5. Kill the operation from the shell using `opid`, which we got earlier:

    ```
    > db.killOp(11587458)
    ```

Administration

How it works...

We will split our explanation into two sections, the first about the current operation details and the second about killing the operation.

The index creation process, in our case, is the long-running operation that we intend to kill. We create a big collection with about 10 million documents, and initiate a background index creation process.

On executing the db.currentOp() operation, we get a document as the result, with an inprog field whose value is an array of other documents, each representing a currently running operation. It is common to get a big list of documents on a busy system. The following is a document taken for the index creation operation:

```
{
  "opid" : 11587458,
  "active" : true,
  "secs_running" : 31,
  "op" : "insert",
  "ns" : "test.system.indexes",
  "insert" : {
    "v" : 1,
    "key" : {
      "i" : 1
    },
    "ns" : "test.currentOpTest",
    "name" : "i_1",
    "background" : 1
  },
  "client" : "127.0.0.1:50895",
  "desc" : "conn10",
  "connectionId" : 10,
  "locks" : {
    "^" : "w",
    "^test" : "W"
  },
  "waitingForLock" : false,
  "msg" : "bg index build Background Index Build Progress: 2214738/10586935 20%",
```

```
"progress" : {
  "done" : 2214740,
  "total" : 10586935
},
"numYields" : 3070,
"lockStats" : {
  "timeLockedMicros" : {
    "r" : NumberLong(0),
    "w" : NumberLong(53831938)
  },
  "timeAcquiringMicros" : {
    "r" : NumberLong(0),
    "w" : NumberLong(31387832)
  }
 }
}
```

We will see what these fields mean in the following table:

Field	Description
opid	This is a unique operation ID identifying the operation. This is the ID to be used to kill an operation.
active	The Boolean value indicates whether the operation has started or not. It is false only if it is waiting to acquire the lock to execute the operation. The value will be `true` once it starts, even if at a point of time where it has yielded the lock and is not executing.
secs_running	This gives the time the operation is executing for in seconds.
op	This indicates the type of the operation. In the case of index creation, it is inserted into a system collection of indexes. The possible values are `insert`, `query`, `getmore`, `update`, `remove`, and `command`.
ns	This is a fully qualified namespace for the target. It will be of the `<database name>.<collection name>` form.
insert	This shows the document that will be inserted in the collection.
query	This is a field that will be present for operations other than the `insert` and `getmore` commands.
client	This is the IP address/hostname and the port of the client who initiated the operation.
desc	This is the description of the client, mostly the client's connection name.

Administration

Field	Description
connectionId	This is the identifier of the client connection from which the request originated.
locks	This is a document containing the locks held for this operation. The document shows the locks held for the operation being analyzed for various databases. The ^ indicates global lock and ^test indicates the lock on the test database. The values here are interesting. The value of ^ is w (lower case). This means that it is not an exclusive write lock, and multiple databases can write concurrently. It is a lock held at the database level. ^test has a value W, which is a global write lock. This means that the write lock on the test database is exclusive and no other operation on any database can occur when this lock is held. The preceding output is for Version 2.4 of Mongo.
waitingForLock	This field indicates whether the operation is waiting for a lock to be acquired. For instance, if the preceding index creation was not a background process, other operations on this database would queue up for the lock to be acquired. This flag for those operations will then be true.
msg	This is a human-readable message for the operation. In this case, we do see a percentage of operation complete, as this is an index creation operation.
progress	This is the state of the operation. The total gives the total number of documents in the collection and done gives the numbers indexed so far. In this case, the collection already had some more documents (over 10 million documents). The percentage of operation completed is computed from these figures.
numYields	This is the number of times the process has yielded the lock to allow other operations to execute. As this is a background index creation process, this number will keep on increasing as the server yields it frequently to let other operations execute. Had it been a foreground process, the lock would never be yielded till the operation completes.
lockStats	This document has more nested documents giving stats of the total time this operation has held the read or write lock, and also the time it waited to acquire the lock. The following are the possible values: ▶ r: This is the time locked for a specific (database level) read lock ▶ w: This is the time locked for a specific (database level) write lock ▶ R: This is the time locked for global read lock ▶ W: This is the time locked for global write lock

If you have a replica set, there will be many more getmore operations on oplog on the primary from secondary.

To see if system operations are executed, we need to pass a `true` value as the parameter to the `currentOp` function call as follows:

> `db.currentOp(true)`

Next, we will see how to kill the user-initiated operation using the `killOp` function. The operation is simply called as follows:

> `db.killOp(<operation id>)`

In our case, the index creation process had the process ID `11587458` and thus it will be killed as follows:

> `db.killOp(11587458)`

On killing any operation, irrespective of whether the given operation ID exists or not, we see the following message on the console:

`{ "info" : "attempting to kill op" }`

Thus, seeing this message doesn't mean that the operation was killed. It just means that the operation, if it exists, will be attempted.

If an operation cannot be killed immediately and if the `killOp` command is issued for it, the `killPending` field in `currentOp` will start appearing for the given operation. For example, execute the following query on the shell:

> `db.currentOpTest.find({$where:'sleep(100000)'})`

This will not return, and the thread executing the query will sleep for 100 seconds. This is an operation that cannot be killed using `killOp`. Try executing `currentOp` from another shell (do not tab for auto completion; your shell may just hang), get the operation ID, and then kill it using the `killOp` command. You should see that the process will still be running if you execute the `currentOp` command, but the document for the process details will now contain a new key, `killPending`, stating that the kill for this operation is requested but pending.

Using profiler to profile operations

In this recipe, we will look at Mongo's in-built profiler that will be used to profile the operations executed on the MongoDB server. It is a utility that is used to log all operations or the slow ones and that can be used to analyze the performance of the server.

Administration

Getting ready

In this recipe, we will be performing some operations on a standalone Mongo instance and profiling them. We need to start a standalone server listening to any port for client connections; in this case, we will stick to the default 27017. If you are not aware of how to start a standalone server, refer to the *Single node installation of MongoDB* recipe in *Chapter 1, Installing and Starting the MongoDB Server*. We also need to start a shell that will be used to perform querying, enable profiling, and view the profiling operation.

How to do it...

1. Once the server is started and the shell is connected to it, execute the following to get the current profiling level:

   ```
   > db.getProfilingLevel()
   ```

   The default level should be 0 (no profiling if we have not set it earlier)

2. Let us set the profiling level to 1 (log slow operations only) and log all the operations slower than 50 ms. Execute the following command on the shell:

   ```
   > db.setProfilingLevel(1, 50)
   ```

3. Now, let us execute an `insert` operation into a collection and then execute a couple of queries:

   ```
   > db.profilingTest.insert({i:1})
   > db.profilingTest.find()
   > db.profilingTest.find({$where:'sleep(70)'})
   ```

4. Now execute the query on the following collection as follows:

   ```
   > db.system.profile.find().pretty()
   ```

How it works...

Profiling is something that will not be enabled by default. If you are happy with the performance of the database, there is no reason to enable the profiler. It is only when one feels that there is some room for improvement and wants to target some expensive operations taking place. An important question is, what classifies an operation to be slow? The answer is, it varies from application to application. By default, in Mongo, slow means any operation above 100 ms. However, while setting the profiling level, you may choose the threshold value.

There are three possible values for profiling levels:

- 0: Disable profiling
- 1: Enable profiling for slow operations where the threshold value for an operation to be classified as "slow" is provided with the call while setting the profiling level
- 2: Profile all operations

While profiling all operations might not be a very good idea and might not commonly be used, as we shall soon see, setting the value to 1 with a threshold provided to it is a good way to monitor slow operations.

If we look at the steps we executed, we see that we can get the current profiling level by executing the db.getProfilingLevel() operation. To get more information, such as what value is set as a threshold for the slow operations, we can execute db.getProfilingStatus(), which returns a document with the profiling level and the threshold value for slow operations.

For setting the profiling level, we call the db.setProfilingLevel() method. In our case, we set it for logging all operations taking more than 50 ms as db.setProfilingLevel(1, 50).

To disable profiling, simply execute db.setProfilingLevel(0).

What we do next is execute three operations; one to insert a document, one to find all documents, and finally, a find that calls sleep with a value of 70 ms to slow it down.

The final step is to see these profiled operations that are logged in the system.profile collection. We execute a find operation to see the operations logged. For my execution, the insert and the final find operation with the sleep were logged.

Obviously, this profiling has some overhead but a negligible one. Hence, we will not enable it by default, but only when we want to profile slow operations. Also, another question is, will this profiling collection increase over a period of time? The answer is no, as this is a capped collection. Capped collections are fixed-size collections, which preserve insertion orders and act as circular queues filling in the new documents and discarding the oldest when it gets full. A query on system.namespaces should show the stats. The query execution will show the following output for the system.profile collection:

```
{"name":"test.system.profile", "options":{"capped":true, "size":1048576
}}
```

As we see, the size of the collection is 1 MB, which is incredibly small. Setting the profiling level to 2 will thus easily overwrite the data on busy systems. One may also choose to explicitly create a collection, with the name system.profile, as a capped collection of any size you prefer, should you choose to retain more operations in it. To create a capped collection explicitly, you may execute the following query from the Mongo shell:

```
> db.createCollection('system.profile', {capped:1, size: 1048576})
```

Administration

Obviously, the size chosen is arbitrary, and you are free to allocate any size to this collection, based on how frequently the data gets filled and how much profiling data you want to keep before it gets overwritten.

As this is a capped collection, and the insertion order is preserved, a query with the sort order `{$natural:-1}` will be perfectly fine and very efficient at finding operations in the reverse order of execution time.

Finally, we will take a look at the document that got inserted in the `system.profile` collection and see which operations it has logged:

```
{
  "op" : "query",
  "ns" : "test.profilingTest",
  "query" : {
    "$where" : "sleep(70)"
  },
  "ntoreturn" : 0,
  "ntoskip" : 0,
  "nscanned" : 1,
  "keyUpdates" : 0,
  "numYield" : 0,
  "lockStats" : {
    "timeLockedMicros" : {
      "r" : NumberLong(188796),
      "w" : NumberLong(0)
    },
    "timeAcquiringMicros" : {
      "r" : NumberLong(5),
      "w" : NumberLong(6)
    }
  },
  "nreturned" : 0,
  "responseLength" : 20,
  "millis" : 188,
  "ts" : ISODate("2014-01-27T17:37:02.482Z"),
  "client" : "127.0.0.1",
  "allUsers" : [ ],
  "user" : ""
}
```

As we see in the preceding document, there are indeed some interesting stats. Let us look at some of them in the following table. Some of these fields are identical to the fields we see when we execute the `db.currentOp()` operation from the shell:

Field	Description
`op`	This is the operation that got executed. In this case, it was a `find` operation and thus, it is a query in this case.
`ns`	This is the fully qualified name of the collection on which the operation was performed. It will be in the `<database>.<collection name>` format.
`query`	This shows the query that got executed on the server.
`nscanned`	This has a similar meaning to explain plan in the relational database. It is the total number of documents and index entries scanned.
`numYields`	This is the number of times the lock was yielded when the operation was executed.
`lockStats`	This has some interesting stats for the time taken to acquire the lock and the time for which the lock was held.
`nreturned`	This is the number of documents returned.
`responseLength`	This is the length of the response in bytes.
`millis`	Most important of all, this is the time taken in milliseconds to execute the operation.
`ts`	This is the time when the operation was executed.
`client`	This is the hostname/IP address of the client who executed the operation.

Setting up users in MongoDB

Security is one of the cornerstones of any enterprise-level system. Not always will you find a system in a completely safe and secure environment to allow unauthenticated user access to it. Apart from test environments, almost every production environment requires proper access rights and perhaps, an audit of the system access too. Mongo security has multiple aspects:

- Access rights for the end users accessing the system. There will be multiple roles, such as admin, read-only users, and read/write nonadministrative users.
- Authentication of the nodes that are added to the replica set. In a replica set, one should only be allowed to add authenticated systems. The integrity of the system will be compromised if any unauthenticated node is added to the replica set.
- Encryption of the data that is transmitted across the wire between the nodes of the replica sets, or even the client and the server (or the mongos process in the case of a sharded setup).

Administration

In this recipe and the next one, we will be looking at how to address the first two points mentioned in the preceding bullet list. The last point, about encrypting the data being transmitted on the wire, is not supported by default by the community edition of Mongo, and it will need a rebuild of the Mongo database with the `ssl` option enabled.

 All the steps are executed on the MongoDB server Version 2.4.6, and all the explanations hold true for this version. There are quite a few changes related to the content we discuss here that are present in Version 2.6 of MongoDB. Any 2.6-specific details will be mentioned as and when needed.

Getting ready

In this recipe, we will be setting up users for a standalone Mongo instance. We need to start a standalone server listening to any port for client connections; in this case, we will stick to the default `27017`. If you are not aware of how to start a standalone server, refer to the *Single node installation of MongoDB* recipe in *Chapter 1, Installing and Starting the MongoDB Server*. We also need to start a shell that will be used for this admin operation. For a replica set, we will only be connected to a primary instance and will perform these operations.

How to do it...

We will add an admin user, a read-only user for the `test` database, and a read-write user for the `test` database in this recipe.

The following are assumed at this point:

- The server is started, up and running and we are connected to it from the shell.
- The server is started without any special command-line argument other than those mentioned in *Chapter 1, Installing and Starting the MongoDB Server*, for starting a single node. Thus, we have full access to the server for any user.

Let's get started:

1. The first step is to create an admin user. Note that, till Version 2.4 of MongoDB, the method name is `addUser`. However, in Version 2.6 of MongoDB, the method is `createUser`. We will look at both methods of creating the users. Execute steps 3 and 4 if you are working on a MongoDB server, Version 2.4 and steps 5 and 6 if you are working on a MongoDB server, Version 2.6.
2. Execute the following command in the Mongo shell to switch to the `admin` database:

    ```
    > use admin
    ```

3. In the `admin` database, we will add a user called `admin` and the password as `admin`:

```
> db.addUser('admin', 'admin')
{
    "user" : "admin",
    "readOnly" : false,
    "pwd" : "7c67ef13bbd4cae106d959320af3f704",
    "_id" : ObjectId("52ea98ef2d00f6e6fb1fcdba")
}
```

4. We will now switch to the `test` database as follows:

```
> use test
```

5. In the `test` database, we will create two users, a read-only user called `read_user` and a read/write user called `write_user`. The password for both these users is the same as their usernames.

6. Execute the following commands to create these users:

```
> db.addUser({user:'read_user', pwd:'read_user', roles:['read']})
{
    "user" : "read_user",
    "pwd" : "60477dd7460977860674077dc0039102",
    "roles" : [
        "read"
    ],
    "_id" : ObjectId("52ee29012d00f6e6fb1fcdbc")
}
> db.addUser({user:'write_user', pwd:'write_user', roles:['readWrite']})
{
    "user" : "write_user",
    "pwd" : "7944cf3480b0eabbf0cff4498ed9652b",
    "roles" : [
        "readWrite"
    ],
    "_id" : ObjectId("52ee292c2d00f6e6fb1fcdbd")
}
```

Administration

7. We will look at how to create users in the `admin` and `test` databases in Version 2.6 of MongoDB. These steps are identical to Version 2.4 of MongoDB, except for the name of the methods. There are some additional features for this method that we will see in detail in the next section. First, we start by creating the admin user in the `admin` database, as follows:

   ```
   > use admin
   > db.createUser({
     user:'admin', pwd:'admin',
     customData:{desc:'The admin user for admin db'},
     roles:['readWrite', 'dbAdmin', 'clusterAdmin']
     }
   )
   ```

8. We will add `read_user` and `write_user` to the `test` database. To add the users, execute the following commands from the Mongo shell:

   ```
   > use test
   > db.createUser({
     user:'read_user', pwd:'read_user',
     customData:{desc:'The read only user for test database'},
     roles:['read']
     }
   )
   > db.createUser({
     user:'write_user', pwd:'write_user',
     customData:{desc:'The read/write user for test database'},
     roles:['readWrite']
     }
   )
   ```

9. Now shut down the MongoDB server and the close the shell too. Restart the MongoDB server but with the `--auth` option on the command line, as follows:

   ```
   $ mongod .. <other options as provided earlier> --auth
   ```

10. Now connect to the server from the newly opened Mongo shell and execute the following command:

    ```
    > db.testAuth.find()
    ```

    The `testAuth` collection need not exist, but you should see an error stating that we are not authorized to query the collection

Chapter 4

11. We will now log in from the shell using `read_user` as follows:

    > `db.auth('read_user', 'read_user')`

12. We will now execute the same `find` operation as follows (note that the `find` operation should not give an error and may not return any results, depending on whether the collection exists or not):

    > `db.testAuth.find()`

13. Now we will try to insert a document as follows (note that we should get an error stating that you are not authorized to insert data in this collection):

    > `db.testAuth.insert({i:1})`

14. We will now log out and log in again, but with a write user as follows. Note the difference in the way we log in this time around, as against the previous instance. We are providing a document as the parameter to the `auth` function, whereas, in the previous case, we passed two parameters for the username and password.

    > `db.logout()`

    > `db.auth({user:'write_user', pwd:'write_user'})`

15. Now execute the `insert` operation again as follows (this time, it should work):

    > `db.testAuth.insert({i:1})`

16. Now execute the following command on the shell. You should get the unauthorized error:

    > `db.serverStatus()`

17. We will now switch to the `admin` database. We are currently connected to the server using `write_user`, which has read/write permissions on the `test` database. From the Mongo shell, try to execute the following commands:

    > `use admin`

    > `show collections`

18. Close the Mongo shell or open a new shell, as follows, from the operating system's console. This should take us directly to the `admin` database:

    `$ mongo -u admin -p admin admin`

19. Now execute the following on the shell. It should show us the collections in the `admin` database:

    > `show collections`

20. Try and execute the following operation:

    > `db.serverStatus()`

    Execute this step if you are on Version 2.4 of MongoDB and create the admin user using the `db.addUser('<user name>', '<password>')`.

Administration

21. Switch to the `test` database and execute the `insert` and `find` operations as follows:

    ```
    > use test
    > db.testAuth.insert({i:1})
    > db.testAuth.find()
    ```

How it works...

We executed a lot of steps and now we will take a closer look at them.

Initially, the server is started without the `--auth` option; hence, no security is enforced by default.

Version 2.4 of MongoDB is where we create a user in the `admin` database using the `addUser(<userName>, <password>)` form of the method. This creates a user in the `admin` database; this special user has read/write access to all the databases and can run admin commands, such as `db.serverStatus()`, and other replication and sharding-related commands. All users created in databases other than admin, whether read or write, will only be able to access the collections in their respective databases.

In version 2.6, however, we create the admin user using the `db.createUser` method. Let us take a closer look at this method first. The signature of the method to create the user is `createUser(user, writeConcern)`. The first parameter is the user, which actually is a JSON document, and the second parameter is the write concern to use for user creation. The JSON document for the user has the following format:

```
{
  'user' : <user name>,
  'pwd' : <password>,
  'customData': {<JSON document providing any user specific data>}
  'roles':[<roles of the user>]
}
```

The roles provided here can be provided as follows, assuming that the current database when the user is created is `test` on the shell:

```
[{'role' : 'read', 'db':'reports'}, 'readWrite']
```

This gives the user that is being created read access to the `db` reports and `readWrite` access to the `test` database. Let us see the complete user creation call for the test user:

```
> use test
> db.createUser({
  user:'test', pwd:'test',
```

```
    customData:{desc:'read access on reports and readWrite access on
test'},
    roles:[
      {role:'read', db : 'reports'},
      'readWrite'
    ]
  }
)
```

The write concern, which is an optional parameter, can be provided as the JSON document. Some sample values are `{w:1}` and `{w:'majority'}`.

Coming back to the admin user creation, we created the user in step 4 using the `createUser` method and gave three roles to this user in the `admin` database.

In steps 4 and 6, we created the read and read/write users in the `test` database using the `addUser` method for version 2.4 and the `createUser` method for version 2.6. The JSON document for the creation of a user in version 2.4 is identical to the user JSON document in version 2.6, except for a couple of differences. First, there is no `customData` field and second, the `roles` array contains string values only for the user roles.

The JSON document for the user in version 2.4 has the following format:

```
{
    'user' : <user name>,
    'pwd' : <password>,
    'roles':[<string values for roles of the user>]
}
```

We shut down the MongoDB server after the admin read and read-write user creation, and restart it with the `--auth` option.

On starting the server again, we connect to it from the shell, which is in step 9, but unauthenticated. Here, we try to execute a `find` query on a collection in the `test` database; this fails, as we are unauthenticated. This shows that the server now requires appropriate credentials to execute operations on it. In steps 10 to 11, we log in using `read_user` and try to execute a `find` operation first, which succeeds, and then an `insert` operation, which doesn't, as the user has read privileges only. The way to authenticate a user is by invoking `db.auth(<user name>, <password>)` from the shell, and `db.logout()` will log out the current logged in user.

Administration

In steps 13 to 15, we demonstrate that we can perform `insert` operations using `write_user`, but admin operations such as `db.serverStatus()` cannot be executed as these operations execute `adminCommand` on the server. This means that a non admin user is not permitted to invoke these operations. Similarly, when we change the database to admin, the `write_user`, which is from the `test` database, is not permitted to perform any operations such as getting a list of collections or any operation to query a collection in the `admin` database.

In steps 16 to 19, we log in to the shell using the admin user to the `admin` database. Previously, we logged in to the database using the `auth` method; in this case, we used the `-u` and `-p` options to provide the username and the password. We also provided the name of the database to connect to, which is `admin` in this case. Here, we are able to view the collections on the `admin` database and also execute admin operations such as getting the server status. In version 2.6, executing the `db.serverStatus` call is possible, as the user is given the `clusterAdmin` role.

In step 18, we are able to switch to any other database and execute read/write operations. This is a special privilege for the users of the `admin` database, which no other user has. This is possible because we created the user in version 2.4 using the version 2.2 style of user creation, `db.addUser(<username>, <password>)`. The admin user created in version 2.6 is not able to query the `test` database as it would need appropriate read and read/write privileges on the respective databases to perform these operations.

One final thing to note; apart from writing to a collection, a user with write privileges can also create indexes on the collection in which he has write access.

There's more...

In this recipe, we saw how we can create different users and what permissions they have, restricting some sets of operations. In the next recipe, we will see how we can have authentication done at the process level. That is, how one Mongo instance can authenticate itself for being added to a replica set.

See also

- `http://docs.mongodb.org/manual/reference/built-in-roles/` to get details of various in-built roles
- `http://docs.mongodb.org/manual/core/authorization/#user-defined-roles` to learn more about defining custom user roles

Understanding interprocess security in MongoDB

In the previous recipe, we saw how authentication can be enforced for a user to be logged in before allowing any operations on Mongo. In this recipe, we will look at interprocess security. By the term interprocess security, we don't mean encrypting the communication but only ensuring that the node, which is added to a replica set, is authenticated before being added to the replica set.

Getting ready

In this recipe, we will be starting multiple Mongo instances as part of a replica set. Thus, you might have to refer to the *Starting multiple instances as part of a replica set* recipe in *Chapter 1, Installing and Starting the MongoDB Server*, if you are not aware of how to start a replica set. Apart from that, in this recipe, all we will be looking at is how to generate a key file to be used, and the behavior when an unauthenticated node is added to the replica set.

How to do it...

To set the ground, we will be starting three instances, each listening to ports 27000, 27001, and 27002 respectively. The first two will be started by providing them with a path to the key file while the third will not receive this. Later, we will try adding these three instances to the same replica set. Let us take a look at the steps in detail:

1. Let us generate the key file first. There is nothing spectacular about generating the key file. This is as simple as having a file with 6 to 1024 characters from the `base64` character set. On the Linux filesystem, you may choose to generate pseudo random bytes using `openssl`, and encode them to `base64`. The following command will generate 500 random bytes, and these bytes will then be `base64` encoded and written to `keyfile`:

    ```
    $ openssl rand -base64 500 > keyfile
    ```

2. On a Unix filesystem, the key file should not have permissions for world and group, and thus, after it is created, we should execute the following command:

    ```
    $ chmod 400 keyfile
    ```

3. Not giving write permission to the creator ensures that we don't overwrite the contents accidentally. On a Windows platform, however, `openssl` doesn't come out of the box and thus, you have to download it. The archive is extracted and the `bin` folder is added to the operating system's path variable. For Windows, we can download `openssl` from http://gnuwin32.sourceforge.net/packages/openssl.htm.

Administration

4. You may even choose not to generate the key file using the approach mentioned earlier (that is, using `openssl`) and can take an easy way out by just typing plain text in the key file from any text editor of your choice. However, note that the characters `\r`, `\n`, and spaces are stripped off by Mongo and the remaining text is considered as the key. For example, we may create a file with the following content added to the key file. Again, the file will be named `keyfile`:

   `somecontentaddedtothekeyfilefromtheeditorwithoutspaces`

   Using any approach mentioned earlier, we would now have a keyfile in place that will be used for the next steps of the recipe

5. We will now secure the Mongo processes by starting the Mongo instance as follows. I will be starting the Mongo instances on Windows; my key file ID named `keyfile` is placed on `c:\MongoDB`, and the data paths are `c:\MongoDB\data\c1, c:\MongoDB\data\c2`, and `c:\MongoDB\data\c3` respectively, for the three instances.

6. Start the first instance listening to port `27000` as follows:

   ```
   C:\> mongod --dbpath c:\MongoDB\data\c1 --port 27000 --auth
   --keyFile c:\MongoDB\keyfile --replSet secureSet --smallfiles
   --oplogSize 100
   ```

7. Similarly, start the second server listening to port `27001` as follows:

   ```
   C:\> mongod --dbpath c:\MongoDB\data\c2 --port 27001 --auth
   --keyFile c:\MongoDB\keyfile --replSet secureSet --smallfiles
   --oplogSize 100
   ```

8. The third instance will be started, but without the `--auth` and `--keyFile` options, listening to port `27002` as follows:

   ```
   C:\> mongod --dbpath c:\MongoDB\data\c3 --port 27002 --replSet
   secureSet --smallfiles --oplogSize 100
   ```

9. We then start a Mongo shell and connect it to port `27000`, which is the first instance started. From the Mongo shell, we type the following command:

   ```
   > rs.initiate()
   ```

10. In a few seconds, the replica set will be initiated with just one instance in it. We will now try to add two new instances to this replica set. First, the one listening on port `27001`, as follows (you will need to add the appropriate hostname; `Amol-PC` is the hostname in my case):

    ```
    > rs.add({_id:1, host:'Amol-PC:27001'})
    ```

11. We will then execute the following command to confirm the status of the newly added instance, by executing `rs.status()`. It should soon come up as a secondary.

12. We will now finally try and add an instance that was started without the `--auth` and `--keyFile` options, as follows:

    ```
    > rs.add({_id:2, host:'Amol-PC:27002'})
    ```

13. This should add the instance to the replica set, but executing `rs.status()` will show the status of the instance as `UNKNOWN`. The server logs for the instance running on `27002` should show some authentication errors as well.

14. We will finally have to restart this instance. However, this time we provide the `--auth` and `--keyFile` options as follows:

    ```
    C:\> mongod --dbpath c:\MongoDB\data\c3 --port 27002 --replSet secureSet --smallfiles --oplogSize 100 --auth --keyFile c:\MongoDB\keyfile
    ```

15. Once the server is started, connect to it from the shell again and type in `rs.status()`. In a few moments, it should come up as a secondary instance.

There's more...

In this recipe, we explored interprocess security to prevent unauthenticated nodes from being added to the mongo replica set. We still haven't encrypted the data that is being sent over the wire to ensure it's delivered securely. In *Appendix, Concepts for Reference*, we will see how to build the MongoDB server from the source and how to enable encryption of the contents over the wire.

Modifying collection behavior using the collMod command

This is a command that will be executed to change the behavior of a collection in Mongo. It can be thought of as a collection-modifying operation (it is not mentioned anywhere officially though).

For a part of this recipe, knowing about TTL indexes is required.

Getting ready

In this recipe, we will be executing the `collMod` operation on a collection. We need to start a standalone server listening to any port for client connections; in this case, we will stick to the default `27017`. If you are not aware of how to start a standalone server, refer to the *Single node installation of MongoDB* recipe in *Chapter 1, Installing and Starting the MongoDB Server*. We also need to start a shell that will be used for this administration. It is highly recommended you take a look at the *Expiring documents after a fixed interval using the TTL index* and *Expiring documents at a given time using the TTL index* recipes in *Chapter 2, Command-line Operations and Indexes*, if you are not aware of them.

Administration

How to do it...

The `collMod` operation can be used to do a few things:

1. Let us change the space allocation on the disk for the new document being added. A collection needs to exist to execute the `collMod` command. You can try to execute this command against any existing collection. In our case, I am assuming we have a collection in place called `powerOfTwoCol`, which was created using the following command from the Mongo shell:

    ```
    > db.createCollection('powerOfTwoCol')
    ```

2. Once the collection is in place/created, execute the following command:

    ```
    > db.runCommand({collMod: 'powerOfTwoCol', usePowerOf2Sizes: 1})
    ```

3. Let us now change the settings of the TTL index. Assuming we have a collection with a TTL index, as we saw in *Chapter 2, Command-line Operations and Indexes*, we can do that by executing the following command:

    ```
    > db.ttlTest.getIndexes()
    ```

4. To change the expiry time to 800 ms from 300 ms, execute the following command:

    ```
    > db.runCommand({collMod: 'ttlTest', index: {keyPattern: {createDate:1}, expireAfterSeconds:800}})
    ```

How it works...

The `collMod` command always has the `{collMod : <name of the collection>, <collmod operation>}` format. There are two possible operations currently supported that we will see. We will break our explanation into two parts.

First, we will see what happens by setting `usePowerOf2Sizes`. If a collection is heavy on updates and the documents grow in size, it will be moved on the disk when it can no longer grow where it is placed. This causes a hole to be left on the disk space for the collection at the place where the document originally was. Mongo uses these holes to accommodate new documents wherever possible. However, by using the `usePowerOf2Sizes` setting, Mongo allocates disk space in numbers by the power of two (32, 64, 128, 256, ...), with the minimum value being 32. This setting does use a few more extra spaces as compared to a normal document without this setting, as the disk space used is rounded always by the power of two. However, in the long term, when the documents get updated frequently and grow in size, the disk usage is better, so is the performance, as document movement is reduced. Thus, if you foresee this pattern of documents growing in size with time, setting this option might be a good idea. However, for patterns where documents are just inserted and rarely updated, we are better off with the default settings (till version 2.4). Also, if the collection already has data when this option is set, the subsequent allocation for the new documents would be by the power of two, without affecting the existing documents.

From Version 2.6 of MongoDB, the `usePowerOf2Sizes` strategy is the default option for all collections and thus, `usePowerOf2Sizes:false` is the only sensible option to use in the `collMod` operation. When starting the server, a new server startup parameter `newCollectionUsePowerOf2Sizes` is available and defaults to the value `true`. This option can be used to disable the `usePowerOf2Sizes` setting by providing the value `false` to it. Setting this value to `false` will ensure that the size allocated to a new document will use the strategy that is followed till version 2.4 by default, which provides space that is needed by the document times the padding factor.

The second operation by using `collMod` is to change the TTL index. If a TTL index has already been created and the time to live needs to be changed after creation, we use the `collMod` command. This operation-specific field is as follows:

```
{index: {keyPattern: <the field on which the index was originally created>, expireAfterSeconds:<new time to be used for TTL of the index>}}
```

The `keyPattern` is the field on which the TTL index is created, and `expireAfterSeconds` will contain the new time to be changed to. On successful execution, we should see the following output in the shell:

```
{ "expireAfterSeconds_old" : 300, "expireAfterSeconds_new" : 800, "ok" : 1 }
```

Setting up MongoDB as a Windows Service

Windows Services are long-running applications that run in the background, just like daemon threads. Databases are good candidates for such services, whereby they start and stop when the host machines start and stop (you may, however, choose to manually start/stop a service). Many database vendors do provide a feature to start the database as a service, when installed on the server. MongoDB also lets you do that, and that is what we will see in this recipe.

Getting ready

Refer to the *Single node installation of MongoDB with options from the config file* recipe in *Chapter 1, Installing and Starting the MongoDB Server,* to get information on how to start the MongoDB server using an external configuration file. As in this case, Mongo is run as a service, it cannot be provided with command-line arguments, and configuring it from a configuration file is the only alternative. Refer to the prerequisites of the *Single node installation of MongoDB* recipe in *Chapter 1, Installing and Starting the MongoDB Server.* This is all we will need for this recipe.

Administration

How to do it...

Let's take a look at the steps in detail:

1. We will first create a config file with three configuration values, namely, the port, db path, and logfile path. We name the file `mongo.conf` and keep it in `c:\conf\mongo.conf` with the following three entries in it (you may choose any path for the config file location, database, and logs):

   ```
   port = 27000
   dbpath = c:\data\mongo\db
   logpath = c:\logs\mongo.log
   ```

2. Execute the following steps from the Windows terminal, which you may need to execute as an administrator. In Windows 7, execute the following steps:

   1. Press the Windows key on your keyboard.
   2. In the **Search programs and files** space, type `cmd`.
   3. In the programs, the command prompt program will be seen. Right-click on it and select **Run as administrator**.

3. In the shell, execute the following command:

   ```
   C:\> mongod --config c:\conf\mongo.conf -install
   ```

   The log printed out on the console should confirm that the service is installed properly

4. The service can be started from the console as follows:

   ```
   C:\> net start MongoDB
   ```

5. The service can be stopped as follows:

   ```
   C:\> net stop MongoDB
   ```

6. Type `services.msc` in the **Run** window (Windows button + R). In the opened management console, search for the MongoDB service. We should see it as follows:

	Name	Description	Status	Startup Type	Log On As
Services (Local)					
Mongo DB	McAfee Security S...	McAfee Sec...		Manual	Local Syste..
Start the service	Media Center Exte...	Allows Med...		Disabled	Local Servic
	Microsoft .NET Fr...	Microsoft		Manual	Local Syste..
Description:	Microsoft .NET Fr...	Microsoft		Manual	Local Syste..
Mongo DB Server	Microsoft iSCSI Ini...	Manages In...		Manual	Local Syste..
	Microsoft SharePo...			Manual	Local Servic
	Microsoft Softwar...	Manages so...		Manual	Local Syste..
	Mongo DB	Mongo DB ...		Automatic	Local Syste..

7. The service is automatic, that is, it will be started when the operating system starts. It can be changed to manual by right-clicking on the service and clicking on **properties**.
8. To remove a service, we need to execute the following command from the command prompt:

    `C:\>mongod --remove`

9. There are more options available that can be used to configure the name of the service, display name, description, and the user account that is used to run the service. These can be provided as command-line arguments. Execute the following command to see the possible options, and take a look at the **Windows Service Control Manager** options:

    `C:\> mongod --help`

Configuring a replica set

We have had a good discussion on what a replica set is and how to start a simple replica set, in the *Starting multiple instances as part of a replica set* recipe in *Chapter 1, Installing and Starting the MongoDB Server*. In the *Understanding interprocess security in MongoDB* recipe, we saw how to start a replica set with interprocess authentication. To be honest, that is pretty much what we do while setting up a standard replica set. However, there are a few configurations that one must know; one must also be aware of how they affect the replica set's behavior. Note that we are still not discussing tag aware replication in this recipe; it will be taken up later in this chapter as a separate recipe *Building tagged replica sets*.

Getting ready

Refer to the recipe *Starting multiple instances as part of a replica set* in *Chapter 1, Installing and Starting the MongoDB Server*, for the prerequisites and to know about the replica set basics. Go ahead and set up a simple three-node replica set on your computer, as mentioned in the recipe.

Before we go ahead with the configurations, we will see what elections are in a replica set and how they work from a high level. It is good to know about elections because some of the configuration options affect the voting process in the elections.

Elections in a replica set

A Mongo replica set has one primary instance and multiple secondary instances. All writes happen only through the primary instance and are replicated to the secondary instances. Read operations can happen from secondary instances, depending on the read preference. Refer to the *Read preference for querying* section in *Appendix, Concepts for Reference*, to know what read preference is. However, if the primary goes down or is not reachable for some reason, the replica set becomes unavailable for writes. A Mongo replica set has a feature to automatically failover to a secondary, by promoting it to a primary and making the set available to clients for both read and write operations. The replica set remains unavailable for that brief moment till a new primary comes up.

All this sounds good, but the question is, who decides what the new primary instance will be? The process of choosing a new primary happens through an election. Whenever any secondary detects that it cannot reach out to a primary, it asks all the replica set nodes in the instance to elect themselves as the new primary.

All other nodes in the replica set that receive this request for the election of the primary will perform certain checks before they vote a *yes* to the secondary requesting an election. Let's take a look at the steps:

1. They will first check whether the existing primary is reachable. This is necessary because the secondary requesting the re-election is not able to reach the primary, possibly because of a network partition, in which case it should not be allowed to become a primary. In such a case, the instance receiving the request will vote a *no*.

2. Secondly, the instance will check the state of replicating itself with the secondary requesting the election. If it finds that the requesting secondary is behind itself in the replicated data, it would vote a *no*.

3. Finally, the primary is not reachable, but some instance with higher priority than the secondary requesting the re-election is reachable from it. This again is possible if the secondary requesting the re-election can't reach out to the secondary with higher priority, possibly due to a network partition. In this scenario, the instance receiving the request for election will vote a *no*.

The preceding checks are pretty much what will be happening (not necessarily in the order mentioned here) during the re-election. If these checks pass, the instance votes a *yes*.

The election is void if even a single instance votes *no*. However, if none of the instances have voted *no*, then the secondary that requests the election will become a new primary if it receives a *yes* from the majority of instances. If the election becomes void, there will be a re-election with the same secondary or any other instance requesting an election with the preceding mentioned process, till a new primary is elected.

Now that we have an idea about elections in a replica set and the terminologies, let us look at some replica set configurations. A few of these options are related to votes, and we start by looking at these options first.

Basic configuration for a replica set

From *Chapter 1*, *Installing and Starting the MongoDB Server*, when we set up a replica set, we have a configuration similar to the following one. The basic replica set configuration for a three-member set is as follows:

```
{
  "_id" : "replSet",
  "members" : [
    {
      "_id" : 0,
      "host" : "Amol-PC:27000"
    },
    {
      "_id" : 1,
      "host" : "Amol-PC:27001"
    },
    {
      "_id" : 2,
      "host" : "Amol-PC:27002"
    }
  ]
}
```

We will not be repeating the entire configuration in the steps in the following sections. All the flags we mention will be added to the document of a particular member in the members array. For example, in the preceding example, if a node with _id as 2 is to be made an arbiter, we will have the following configuration for it in the configuration document shown earlier:

```
{
  "_id" : 2,
  "host" : "Amol-PC:27002"
  "arbiterOnly" : true
}
```

Generally, the steps to reconfigure a replica set that has already been set up are as follows:

1. Assign the configuration document to a variable. If the replica set is already configured, it can be obtained using the rs.conf() call from the shell as follows:

    ```
    > var conf = rs.conf()
    ```

Administration

2. The `members` field in the document is an array of documents for each individual member of a replica set. To add a new property to a particular member, we need to execute the following command. For instance, if we want to add the `votes` key and set its value to 2 for the third member of the replica set (`index` 2 in the array), we execute the following command:

   ```
   > conf.members[2].votes = 2
   ```

3. Just changing the JSON document won't change the replica set. We need to reconfigure it as follows if the replica set is already in place:

   ```
   > rs.reconfig(conf)
   ```

4. If the configuration is done for the first time, we will call the following command:

   ```
   > rs.initiate (conf)
   ```

For all the steps given in the next section, you need to follow the preceding steps to reconfigure or initiate the replica set, unless some other steps are mentioned explicitly.

How to do it...

In this recipe, we will look at some of the possible configurations that can be used in a replica set. The explanation here will be minimal with all the explanations done as usual in the next section.

1. The first configuration is an arbiter option that is used to configure a replica set member as a member that holds no data but only has rights to vote. The following key needs to be added to the configuration of the member who will be made an arbiter:

   ```
   {_id: ... , 'arbiterOnly': true }
   ```

2. One thing to remember regarding this configuration is that once a replica set is initiated, no existing member can be changed to an arbiter from a nonarbiter node and vice versa. However, we can add an arbiter to an existing replica set using the helper function `rs.addArb(<hostname>:<port>)`. For example, to add an arbiter listening to port 27004 to an existing replica set, the following command was executed on my machine:

   ```
   > rs.addArb('Amol-PC:27004')
   ```

   When the server starts to listen to port 27004, and `rs.status()` is executed from the Mongo shell, we see that `state` and `strState` for this member are 7 and `ARBITER` respectively.

3. The next option, `votes`, affects the number of votes a member gets in the election. By default, all members get one vote each. This option can be used to change the number of votes a particular member gets. It can be set as follows:

   ```
   {_id: ... , 'votes': <number of votes>}
   ```

Chapter 4

The votes of existing members of a replica set can be changed and the replica set can be reconfigured using rs.reconfig().

Though the option votes is available, which can potentially change the number of votes to form a majority, it usually doesn't add much value and is not a recommended option to use in production.

4. The next replica set configuration option is called priority. It determines the eligibility of a replica set member to become a primary (or not to become a primary). The option is set as follows:

   `{_id: ... , 'priority': <priority number>}`

5. A higher number indicates more likelihood of becoming a primary. The primary will always be the one with the highest priority among the members alive in a replica set. Setting this option in an already configured replica set will trigger an election.

6. Setting the priority option to 0 will ensure that a member will never become a primary.

7. The next option we look at is hidden. Setting the value of this option to true ensures that the replica set member is hidden. The option is set as follows:

   `{_id: ... , 'hidden': <true/false>}`

   One thing to keep in mind is that, when a replica set member is hidden, its priority too should be made 0 to ensure it doesn't become primary. Though this seems redundant, as of the current version, the value or priority needs to be set explicitly.

8. When a programming language client connects to a replica set, it will not be able to discover hidden members. However, after executing rs.status() from the shell, the member's status would be visible.

9. The next option we will look at is the slaveDelay option. This option is used to set the lag in time for the slave from the primary of the replica set. The option is set as follows:

   `{_id: ... , 'slaveDelay': <number of seconds to lag>}`

10. Like the hidden member, slave delayed members too should have the priority option set to 0 to ensure they don't ever become primary. This needs to be set explicitly.

11. The final configuration option we will be looking at is buildIndexes. This value if not specified. By default, the value is true, which indicates that if an index is created on the primary, it needs to be replicated on the secondary too. The option is set as follows:

    `{_id: ... , 'buildIndexes': <true/false>}`

12. If the value of buildIndexes is set to false, the priority is set to 0 to ensure they don't ever become primary. This needs to be set explicitly. Also, this option cannot be set after the replica set is initiated. Just like an arbiter node, this needs to be set when the replica set is being created or when a new member node is being added to the replica set.

How it works...

In this section, we will explain and understand the significance of different types of members and the configuration options we saw in the previous section.

A replica set member as an arbiter

The English meaning of the word "arbiter" is a judge who resolves a dispute. In the case of replica sets, the arbiter node is present just to vote in the case of elections and not to replicate any data. This is, in fact, a pretty common scenario due to the fact that that a Mongo replica set needs to have at least three instances (and preferably an odd number of instances, three or more). A lot of applications do not need to maintain three copies of data and are happy with just two instances, one primary and a secondary with the data.

Consider the scenario where only two instances are present in the replica set. When the primary goes down, the secondary instance cannot form a proper majority because it only has 50 percent of the votes (its own votes) and thus, it cannot become a primary. If a majority of the secondary instances go down, then the primary instance steps down from the primary and becomes a secondary, thus making the replica set unavailable for writes. Thus, a two-node replica set is useless, as it doesn't stay available even when any of the instances go down. It defeats the purpose of setting up a replica set and thus, a minimum of three instances are needed in a replica set.

Arbiters come in handy in such scenarios. We set up a replica set instance with three instances, with only two having data and one acting as an arbiter. We need not maintain three copies of data at the same time; we eliminate the problem we face, by setting up a two-instance replica set.

Priority of replica set members

This is an option whose use is enforced by other options as well, though it can be used on its own in some cases. The options that enforce its usage are `hidden`, `slaveDelay`, and `buildIndexes`, where we don't want the member with one of these three options to ever be made primary. We will look at these options soon.

Some more possible use cases, where we never want a replica set to become a primary, are as follows:

- When the hardware configuration of a member is not able to deal with the write and read requests, should it become a primary; and the only reason it is being put in there is for replicating the data.

- We have a multi data center setup, where one replica set instance is present in another data center for the sake of geographically distributing the data for disaster recovery purposes. Ideally, the network latency between the application server hosting the application and the database should be minimal for optimum performance. This can be achieved if both the servers (the application server and database server) are in the same data center. Not changing the priority of the replica set instance in another data center makes it equally eligible for being chosen as a primary, thus compromising the application's performance if the server from another data center gets chosen as the primary. In such scenarios, we can set the priority to be 0 for the server in the second data center, and a manual cutover will be needed by the administrator to fail over to another data center, should an emergency arise.

In both these scenarios, we can also have the respective members hidden so that the application client doesn't have a view of these members in the first place.

Just as we set the priority to 0 to not allow one to be the primary, we can also be biased towards one member being the primary, whenever it is available, by setting its priority to a value greater than one, because the default value of the `priority` field is 1.

Suppose we have a scenario where, for budget reasons, we have one of the members storing data on SSDs and the remaining data on spinning disks. We will ideally want the member with SSDs to be the primary, whenever the primary server is up and running. It is only when it is not available that we will want another member to become a primary. In such scenarios, we can set the priority of the member running on SSD to a value greater than 1. The value doesn't really matter as long as it is greater than the rest; that is, setting it to 1.5 or 2 makes no difference as long as the priority of the other members is less.

Hidden, votes, slave delayed, and build index configurations

The term hidden for a replica set node is for an application client that is connected to the replica set and not for an administrator. For an administrator, it is equally important for the hidden members to be monitored and thus, their state is seen in the `rs.status()` response. Hidden members participate in elections too, just like all other members.

Though `votes` is an option that is not a recommended solution to a problem, there is an interesting behavior that needs to be mentioned. Suppose you have a three-member replica set. With each instance of the replica set having one vote by default, we have a total of three votes in the replica set. For a replica set to allow writes, a majority of voting members should be up. However, the calculation of a majority doesn't happen using the number of members up but by the total number of votes. Let us see how.

Administration

By default, with one vote each, if one of the members is down, we have two out of a total of three votes available, and thus, the replica set continues to operate. However, if we have one member with the number of votes set to 2, we now have a total of four votes (*1 + 1 + 2*) in the replica set. If this member goes down, even though it is secondary, the primary will automatically step down, and the replica set will be left with no primary, thus not allowing writes. This happens because two out of four possible votes are now gone and we no longer have a majority of the votes available. If this member with two votes is a primary, then again no majority can be formed as there are just a maximum of two votes out of four available, and a primary won't be elected. Thus in general, as a rule of thumb, if you are tempted to use this votes configuration option for your use case, think again, as you may very well use other options such as `priority` and `arbiterOnly` to address these use cases.

From Version 2.6 of MongoDB, the `votes` option is deprecated, and the following message gets printed in the logs:

```
[rsMgr]    WARNING: Having more than 1 vote on a single replicaset member is
[rsMgr]    deprecated, as it causes issues with majority write concern. For
[rsMgr]    more information, see http://dochub.mongodb.org/core/replica-set-votes-deprecated
```

Thus, it is recommended not to use this option and prefer an alternative configuration option; in some future version of MongoDB, it might not even be supported.

For the `slaveDelay` option, the most common use case is to ensure that the data in a member at a particular point of time lags behind the primary by the provided number of seconds. It can be restored if some unforeseen error happens, say, a human erroneously updating some data. Remember, the longer the time delay, the longer the time we get to recover, but at the cost of possibly stale data.

Finally, we'll see the `buildIndexes` option. This is useful in cases where we have a replica set member with nonproduction standard hardware and the cost of maintaining the indexes is not worth it. You may choose to set this option for members where no queries are executed on them. Obviously, if you set this option, they can never become primary members and thus, the `priority` option is enforced to be set to 0.

There's more...

You can achieve some interesting things using tags in replica sets. This will be discussed in a later recipe after we learn about tags in the *Building tagged replica sets* recipe.

Stepping down as a primary instance from the replica set

There are times when, for maintenance activity during business hours, we need to take a server out from the replica set, perform the maintenance, and put it back in the replica set. If the server to be worked upon is the primary, we somehow need to step down from the primary member position, conduct a re-election, and ensure that it doesn't get re-elected for a minimum given time frame. After the server becomes a secondary once the step down operation is successful, we can take it out of the replica set, perform the maintenance activity, and put it back in the replica set.

Getting ready

Refer to the *Starting multiple instances as part of a replica set* recipe in *Chapter 1, Installing and Starting the MongoDB Server*, for the prerequisites and to know about replica set basics. Set up a simple three-node replica set on your computer as mentioned in the recipe.

How to do it...

Assuming that at this point of time we have a replica set up and running, perform the following steps:

1. Execute the following command from the shell connected to one of the replica set members and see which instance is currently the primary:

    ```
    > rs.status()
    ```

2. Connect to that primary instance from the Mongo shell and execute the following command on the shell:

    ```
    > rs.stepDown()
    ```

3. The shell should reconnect again, and you should see that the instance connected to, which was initially a primary instance, now becomes secondary. Execute the following command from the shell so that a new primary is now re-elected:

    ```
    > rs.status()
    ```

4. You may now connect to the primary, modify the replica set configuration, and go ahead with the administration on the servers.

How it works...

The steps we saw in the previous section are pretty simple, but there are a couple of interesting things that we will see.

Administration

The `rs.stepDown()` method did not have any parameter. The function can in fact take a numeric value, the number of seconds for which the instance stepped down won't participate in the elections and won't become a primary; the default value for this is 60 seconds.

Another interesting thing to try out is, what if the instance that was asked to step down has a higher priority than other instances? Well, it turns out that the priority doesn't matter when you step down. The instance stepped down will not become primary, no matter what, for the provided number of seconds. However, if the priority is set for the instance stepped down, and it is higher than others, then after the time given to step down elapses, an election will happen, and the instance with the higher priority will become primary again.

Exploring the local database of a replica set

In this recipe, we will explore the `local` database from a replica set's perspective. The `local` database may contain collections that are not specific to replica sets, but we will focus only on the replica-set-specific collections and try to take a look at what's in them and what they mean.

Getting ready

Refer to the *Starting multiple instances as part of a replica set* recipe in *Chapter 1, Installing and Starting the MongoDB Server*, for the prerequisites and to know about replica set basics. Go ahead and set up a simple three-node replica set on your computer, as mentioned in the recipe.

How to do it...

1. With the replica set up and running, we need to open a shell connected to the primary. You may randomly connect to any one member, execute `rs.status()`, and then determine the primary.

2. With the shell opened, first switch to the `local` database and then view the collections in the `local` database as follows:

   ```
   > use local
   switched to db local
   > show collections
   ```

3. You should find a collection called `me`. Querying this collection should show us a document that contains the hostname of the server to which we are currently connected:

   ```
   > db.me.findOne()
   ```

   There will be two fields, `hostname` and `_id`. Take note of the `_id` field; it is important.

4. We will now query the `slaves` collection as follows:

   ```
   > db.slaves.find().pretty()
   ```

5. Take a note of the fields present in these documents.

6. The next collection to look at is `replset.minvalid`. You will have to connect to a secondary member from the shell to execute the following query. Switch to the `local` database first as follows:

   ```
   > use local
   switched to db local
   > db.replset.minvalid.find()
   ```

   This collection just contains a single document with a key `ts` and a value, which is the timestamp for the time the secondary we are connected to is synchronized. Note down this time.

7. From the shell in the primary, insert a document in any collection. We will use the database `test`. Execute the following commands from the shell of the primary member:

   ```
   > use test
   switched to db test
   > db.replTest.insert({i:1})
   ```

8. Query the secondary again as follows:

   ```
   > db.replset.minvalid.find()
   ```

   We see that the time against the `ts` field has now incremented corresponding to the time at which this replication happened from primary to secondary. With a slave delayed node, you will see this time getting updated only after the delay period has elapsed.

9. Finally, we will see the `system.replset` collection. This collection is where the replica set configuration is stored. Execute the following command:

   ```
   > db.system.replset.find().pretty()
   ```

   Actually, when we execute `rs.conf()`, the following query gets executed:

   ```
   > db.getSisterDB("local").system.replset.findOne()
   ```

How it works...

The `local` database is a special database that is used to hold the replication and instance-specific details in it. This is a nonreplicated database. Try creating a collection of your own in the `local` database and insert some data in it; it will not be replicated to the secondary nodes.

This database gives us a view of the data stored by Mongo for internal use. However, as an administrator, it is good to know about these collections and the type of data in them.

Most of the collections are pretty straightforward. We will take a closer look at the `slaves` collection. Let's take a look at the following example:

```
{
  "_id" : ObjectId("52f138169da4944dff694e26"),
  "config" : {
    "_id" : 1,
    "host" : "Amol-PC:27001"
  },
  "ns" : "local.oplog.rs",
  "syncedTo" : Timestamp(1391928970, 1)
}
```

This collection contains the document for all the secondary members that have synched from it. The _id field here is not a randomly chosen ID, but has the same value as the _id field of the document in the me collection of the respective secondary member nodes. From the shell of the secondary, execute the `db.me.findOne()` query in the `local` database and we should see that the _id field there should match the _id field of the document present in the `slaves` collection.

The config document we see gives the hostname of the secondary instance that we are referring to. Note that the port and other configuration options of the replica set member are not present in this document. Finally, the `syncedTo` time tells us what time are secondary instances are synced up to with the primary. We saw the `replset.minvalid` collection on the secondary, which tells the time to which it is synced with the primary. This value in `syncedTo` in the primary will be the same as in `replset.minvalid` in the respective secondary.

See also

- The *Understanding and analyzing oplogs* recipe

Chapter 4

Understanding and analyzing oplogs

Oplog is a special collection and forms the backbone of the MongoDB replication. When any write operation or configuration changes are done on the replica set's primary, they are written to the oplog on the primary. All the secondary members then tail this collection to get the changes to be replicated. Tailing is synonymous with the `tail` command in Unix and can only be done on a special type of collection called capped collections. Capped collections are fixed size collections that maintain the insertion order just like a queue. When the collection's allocated space becomes full, the oldest data is overwritten. If you are not aware of capped collections and what tailable cursors are, refer to the *Creating and tailing capped collection cursors in MongoDB* recipe in *Chapter 5, Advanced Operations*, for more details.

Oplog is a capped collection present in the nonreplicated database called `local`. In the previous recipe, we saw what a `local` database is and what collections are present in it. Oplog is something we didn't discuss in the previous recipe, as it demands a lot more explanation and a dedicated recipe is needed to do it justice.

Getting ready

Refer to the *Starting multiple instances as part of a replica set* recipe in *Chapter 1, Installing and Starting the MongoDB Server*, for the prerequisites and to know about the replica set basics. Go ahead and set up a simple three-node replica set on your computer as mentioned in the recipe. Open a shell and connect to the primary member of the replica set. You will need to start the Mongo shell and connect to the primary instance.

How to do it...

1. Execute the following commands after connecting to a primary from the shell to get the timestamp of the last operation present in oplog. We are interested in looking at the operations after this time.

   ```
   > use test
   > local = db.getSisterDB('local')
   > var cutoff = local.oplog.rs.find().sort({ts:-1}).limit(1).next().ts
   ```

2. Execute the following command from the shell. Keep the output in the shell or copy it somewhere. We will analyze it later.

   ```
   > local.system.namespaces.findOne({name:'local.oplog.rs'})
   ```

3. Insert 10 documents as follows:

   ```
   > for(i = 0; i < 10; i++) db.oplogTest.insert({'i':i})
   ```

Administration

4. Execute the following update operation to set a string value for all documents with the value of i greater than 5, which are 6, 7, 8, and 9 in our case. It is a multiupdate operation:

   ```
   > db.oplogTest.update({i:{$gt:5}}, {$set:{val:'str'}}, false, true)
   ```

5. Now create the index as follows:

   ```
   > db.oplogTest.ensureIndex({i:1}, {background:1})
   ```

6. Execute the following query on oplog as follows:

   ```
   > local.oplog.rs.find({ts:{$gt:cutoff}}).pretty()
   ```

How it works...

For those aware of messaging and its terminologies, oplog can be looked at as a topic in the messaging world with one producer, which is the primary instance, and multiple consumers, which are the secondary instances. The primary instance writes to an oplog all the contents that need to be replicated. Thus, any create, update, and delete operations, as well as any reconfigurations on the replica sets will be written to the oplog; and the secondary instances will tail (continuously read the contents of the oplog being added to it, which is similar to a `tail` command with an `-f` option in Unix) the collection to get documents written by the primary. If the secondary has a `slaveDelay` configured, it will not read documents for more than the maximum time minus the `slaveDelay` time from the oplog.

We started by saving an instance of the `local` database in the variable called `local` and identified a cutoff time that we will use to query all the operations we will perform in this recipe from the oplog.

Executing a query on the `system.namespaces` collection in the `local` database shows us that the collection is a capped collection with a fixed size. For performance reasons, capped collections are allocated continuous space on the filesystem and this space is preallocated. The size allocated by the server is dependent on the OS and CPU architecture. While starting the server, the `oplogSize` option can be provided to mention the size of the oplog. The defaults are generally good enough for most cases; however, for development purposes, one may choose to override this value with a smaller value. Oplogs are capped collections that need to be preallocated a space on the disk. This preallocation not only takes time when the replica set is first initialized, but also takes up a fixed amount of disk space. For development purposes, we generally start multiple MongoDB processes as part of the same replica set on the same machine and want them to be up and running as quickly as possible with minimal resource usage. Also, having the entire oplog in memory becomes possible if the oplog size is small. For all these reasons, it is advisable to start local instances for development purposes with a small oplog size.

Chapter 4

We performed some operations, such as insert 10 documents and update four documents, using a multiupdate operation, and created an index. If we query the oplog for entries after the cutoff we computed earlier, we see 10 documents for each insert in it. The document looks as follows:

```
{
        "ts" : Timestamp(1392402144, 1),
        "h" : NumberLong("-4661965417977826137"),
        "v" : 2,
        "op" : "i",
        "ns" : "test.oplogTest",
        "o" : {
                "_id" : ObjectId("52fe5edfd473d2f623718f51"),
                "i" : 0
        }
}
```

As seen in the previous example, we first look at the three fields, namely op, ns, and o. These fields stand for the operation, the fully qualified name of the collection into which the data is being inserted, and the actual object to be inserted. The operation i stands for the insert operation. Note that the value of o, which is the document to be inserted, contains the _id field that got generated on the primary. We should see 10 such documents, one for each insert. What is interesting is to see what happens on a multiupdate operation. The primary puts four documents, one for each of them affected by the updates. In this case, the op value is u, for the update, and the query used to match the document is not the same as we gave in the update function; rather, it is a query that uniquely finds a document based on the _id field. As there is an index already in place for the _id field (created automatically for each collection), this operation to find the document to be updated is not expensive. The value of the o field is the same as the document we passed to the update function from the shell. The sample document in the oplog for the update is as follows:

```
{
    "ts" : Timestamp(1392402620, 1),
    "h" : NumberLong("-7543933489976433166"),
    "v" : 2,
    "op" : "u",
    "ns" : "test.oplogTest",
    "o2" : {
            "_id" : ObjectId("52fe5edfd473d2f623718f57")
    },
    "o" : {
```

Administration

```
            "$set" : {
                "val" : "str"
            }
        }
    }
}
```

The update in the oplog is the same as the one we provided, because the `$set` operation is idempotent, which means you may apply an operation safely any number of times.

However, an update using the `$inc` operator is not idempotent. Let us execute the following `update` query:

```
> db.oplogTest.update({i:9}, {$inc:{i:1}})
```

In this case, the oplog will have the following output as the value of o:

```
"o" : {
    "$set" : {
        "i" : 10
    }
}
```

This nonidempotent operation is put into oplog by Mongo smartly, as an idempotent operation with the value of `i` set to a value that is expected to be after the increment operation once. Thus, it is safe to replay an oplog any number of times without corrupting the data.

Finally, we can see that the index creation process is put in the oplog as an `insert` operation in the `system.indexes` collection. However, there is something to remember during index creation till Version 2.4 of MongoDB. An index creation, whether foreground or background on the primary, is always created in the foreground on a secondary and thus, for that period, replication will not happen on that secondary instance. For large collections, index creation can take hours and thus, the size of the oplog is very important to let the secondary catch up from where it hasn't replicated since the index creation started. However, since version 2.6, index creation initiated in the background on the primary will also be built in the background on secondary instances.

For more details on the index creation on replica sets, visit `http://docs.mongodb.org/master/tutorial/build-indexes-on-replica-sets/`.

Building tagged replica sets

In the *Starting multiple instances as part of a replica set* recipe in *Chapter 1, Installing and Starting the MongoDB Server*, we saw how to set up a simple replica set and what the purpose of a replica set is. We also have a good deal of explanation in *Appendix, Concepts for Reference*, on what write concern is and why it is used. What we saw about write concerns is that they offer a minimum level guarantee for a certain write operation. However, with the concept of tags and write concerns, we can define a variety of rules and conditions that must be satisfied before a write operation is deemed successful and a response is sent to the user.

Consider some common use cases:

- An application wants a write operation to be propagated to at least one server in each of its data centers. This ensures that, in the event of a data center shutdown, other data centers will have the data that was written by the application.

- If there aren't multiple data centers, at least one member of a replica set is kept on a different rack. For instance, if the rack's power supply goes down, the replica set will still be available (not necessarily for writes) as at least one member is running on a different rack. In such scenarios, we would want the write to be propagated to at least two racks before responding to the client with a successful write.

- It is possible that a reporting application queries a group of secondary instances of a replica set to generate some reports regularly (such a secondary might be configured to never become primary). After each write, we want to ensure that the write operation is replicated to at least one reporting replica member, before acknowledging the write as successful.

The preceding use cases are a few of the common use cases that arise and are not addressed using simple write concerns that we have seen earlier. We need a different mechanism to cater to these requirements; replica sets with tags are what we need.

Obviously the next question is, What exactly are tags? Let us take an example of a blog. Various posts in the blog have different tags attached to them. These tags allow us to easily search, group, and relate posts together. Tags are user-defined texts with some meaning attached to it. If we draw an analogy between a blog post and the replica set members, just as we attach tags to a post, we can attach tags to each replica set member. For example, in a multi-data center scenario with two replica set members in data center 1 (dc1) and one member in data center 2 (dc2), we can have the following tags assigned to the members. The name of the key and the value assigned to the tag are arbitrary, and they are chosen during the designing of the application.

Administration

You may even choose to assign any tags, for example, to the administrator who set up the server, if you really find it useful to address your use case.

Replica set member	Tag
Replica set member 1	{'datacentre': 'dc1', 'rack': 'rack-dc1-1'}
Replica set member 2	{'datacentre': 'dc1', 'rack': 'rack-dc1-2'}
Replica set member 3	{'datacentre': 'dc2', 'rack': 'rack-dc2-2'}

This is good enough to lay the foundation of what replica set tags are. In this recipe, we will see how to assign tags to replica set members and, more importantly, how to make use of them to address some of the sample use cases we saw earlier.

Getting ready

Refer to the *Starting multiple instances as part of a replica set* recipe in *Chapter 1, Installing and Starting the MongoDB Server,* for the prerequisites and to know about replica set basics. Go ahead and set up a simple three-node replica set on your computer, as mentioned in the recipe. Open a shell and connect to the primary member of the replica set.

If you need to know about write concerns, refer to the overview of write concerns *Appendix, Concepts for Reference*.

For the purpose of inserting into the database, we will use Python, as it gives us an interactive interface such as the Mongo shell. Refer to the *Installing PyMongo* recipe in *Chapter 3, Programming Language Drivers,* for steps on how to install PyMongo. The Mongo shell would have been the most ideal candidate for the demonstration of the `insert` operations, but there are certain limitations around the usage of the shell with our custom write concern. Technically, any programming language with the write concerns mentioned in the recipe for `insert` operations will work fine.

How to do it...

1. With the replica set started, we will add tags to it and reconfigure using the following commands that are executed from the Mongo shell:

    ```
    > var conf = rs.conf()
    > conf[0].members.tags = {'datacentre': 'dc1', 'rack': 'rack-dc1-1'}
    > conf[1].members.tags = {'datacentre': 'dc1', 'rack': 'rack-dc1-2'}
    > conf[2].members.priority = 0
    > conf[2].members.tags = {'datacentre': 'dc2', 'rack': 'rack-dc2-1'}
    ```

2. With the replica set tags set (note that we have not yet reconfigured the replica set), we need to define some custom write concerns. First, we define one that will ensure that the data gets replicated at least to one server in each data center. Execute the following commands in the Mongo shell again:

   ```
   > conf.settings = {'getLastErrorModes' : {'MultiDC':{datacentre : 2}}}
   > rs.reconfig(conf)
   ```

3. Start the Python shell and execute the following commands:

   ```
   >>> import pymongo
   >>> client = pymongo.MongoReplicaSetClient('localhost:27000,localhost:27001', replicaSet='replSetTest')
   >>> db = client.test
   ```

4. We will now execute the following insert query:

   ```
   >>> db.multiDCTest.insert({'i':1}, w='MultiDC', wtimeout=5000)
   ```

5. The preceding insert query goes through successfully, and ObjectId will be printed out. You may query the collection to confirm from either the Mongo shell or the Python shell.

6. As our primary is one of the servers in data center 1, we will now stop the server listening to port 27002, which is the one with priority 0 and tagged to be in a different data center.

7. Once the server is stopped (you may confirm using the rs.status() helper function from the Mongo shell), execute the following insert query again; this insert should throw an error for timeout:

   ```
   >>> db.multiDCTest.insert({'i':2}, w='MultiDC', wtimeout=5000)
   ```

8. Restart the stopped MongoDB server.

9. Similarly, we can achieve rack awareness by ensuring that the write propagates at least two racks (in any data center) by defining a new configuration from the Mongo shell as follows

   ```
   {'MultiRack':{rack : 2}}
   ```

10. The settings value of the conf object will then be as follows. Once set, reconfigure the replica set again using rs.reconfig(conf) from the Mongo shell as follows:

    ```
    {
        'getLastErrorModes' : {
                'MultiDC':{datacentre : 2},
                'MultiRack':{rack : 2}
        }
    }
    ```

Administration

We saw `WriteConcern` used with replica set tags to achieve functionality such as data center and rack awareness. Let us see how we can use replica set tags with read operations.

11. We will see how to make use of replica set tags with read preference. Let us reconfigure the set by adding one more tag to mark a secondary member that will be used to execute some hourly stats reporting.

12. Execute the following steps to reconfigure the set from the Mongo shell:

    ```
    > var conf = rs.conf()
    > conf.members[2].tags.type = 'reports'
    > rs.reconfig(conf)
    ```

13. This will configure the same member with priority `0` and `1` in a different data center with an additional tag called `type` with the value `reports`.

14. We now go back to the Python shell and execute the following commands:

    ```
    >>> curs = db.multiDCTest.find(read_preference=pymongo.ReadPreference.SECONDARY,
    tag_sets=[{'type':'reports'}])
    >>> curs.next()
    ```

15. The preceding execution should show us one document from the collection (as we had inserted data in this test collection in the previous steps).

16. Stop the instance that we tagged for reporting, that is, the server listening to connections on port `27002`, and execute the following command on the Python shell again:

    ```
    >>> curs = db.multiDCTest.find(read_preference=pymongo.ReadPreference.SECONDARY,
    tag_sets=[{'type':'reports'}])
    >>> curs.next()
    ```

    This time around, the execution should fail and state that no secondary was found with the required tag sets.

How it works...

In this recipe, we did a lot of operations on tagged replica sets and saw how they can affect write operations using `WriteConcern` and read operations using `ReadPreference`. Let us look at them in some detail now.

WriteConcern in tagged replica sets

We set up a replica set that was up and running, which we reconfigured to add tags. We tagged the first two servers in data center 1 and in different racks (with the servers running and listening to ports `27000` and `27001` for client connections), and the third one in data center 2 (with the server listening to port `27002` for client connections). We also ensured that the member in data center 2 doesn't become a primary by setting its priority to 0.

Our first objective is to ensure that write operations to the replica set get replicated to at least one member in the two data centers. To ensure this, we define a write concern as follows:

`{'MultiDC':{datacentre : 2}}`

Here, we first define the name of the write concern as `MultiDC`. The value, which is a JSON object, has one key with the name `datacenter`, which is the same as the key used for the tag we attached to the replica set, and the value is the number 2, which will be looked at as the number of distinct values of the given tag that should acknowledge the write before it is deemed successful.

For instance, in our case, when the write comes to server 1 in data center 1, the number of distinct values of the `datacentre` tag is 1. If the write operation gets replicated to the second server, the number still stays 1, as the value of the `datacentre` tag is the same as the first member. It is only when the third server acknowledges the write operation that the write satisfies the defined condition of replicating the write to two distinct values of the `datacentre` tag in the replica set. Note that the value can only be a number and can not have something such as `{datacentre : 'dc1'}`. This definition is invalid and an error will be thrown while reconfiguring the replica set.

However, we need to register this write concern somewhere with the server. This is done in the final step of the configuration by setting the `settings` value in configuration JSON. The value to set is `getLastErrorModes`. The value of `getLastErrorModes` is a JSON document with all possible write concerns defined in it. We later define one more write concern for writes propagated to at least two racks. This is conceptually in line with the `MultiDC` write concern and thus, we will not be discussing it in detail here. After setting all the required tags and settings, we reconfigure the replica set for the changes to take effect.

Once reconfigured, we perform some write operations using the `MultiDC` write concern. When two members in two distinct data centers are available, the write goes through successfully. However, when the server in the second data center goes down, the write operation times out and throws an exception to the client initiating the write. This demonstrates that the write operation will succeed or fail as per how we intended.

We just saw how these custom tags can be used to address some interesting use cases that are not supported by the product implicitly, as far as write operations are concerned. Similar to write operations, read operations can take full advantage of these tags to address some use cases, such as reading from a fixed set of secondary members that are tagged with a particular value.

Administration

ReadPreference in tagged replica sets

We added another custom tag annotating a member to be used for reporting purposes. We then fired a query operation with the read preference to query a secondary and provided the tag sets that should be looked for before considering the member as a candidate for a read operation. Remember that when using a primary as the read preference, we cannot use tags, and that is the reason we explicitly specified the value of `read_preference` to `SECONDARY`.

Configuring the default shard for nonsharded collections

In the *Starting a simple sharded environment of two shards* recipe in *Chapter 1, Installing and Starting the MongoDB Server*, we set up a simple two-shard server. In the *Connecting to a shard from the Mongo shell and performing operations* recipe in *Chapter 1, Installing and Starting the MongoDB Server*, we added data to a `person` collection that was sharded. However, for any collection that is not sharded, all the documents end up on one shard called the primary shard. This situation is acceptable for small databases with a relatively small number of collections. However, if, the database size increases and at the same time, the number of unsharded collections increase we end up overloading a particular shard (the primary shard for a database) with a lot of data from these unsharded collections. All query operations for such unsharded collections, as well as those on the collections whose particular range in the shard reside on this server instance, will be directed to this. In such a scenario, we can have the primary shard of a database changed to some other instance so that these unsharded collections get balanced out across different instances. In this recipe, we will see how to view this primary shard and change it to some other server whenever needed.

Getting ready

Refer to the *Starting a simple sharded environment of two shards* recipe in *Chapter 1, Installing and Starting the MongoDB Server*, to set up and start a sharded environment. From the shell, connect to the started `mongos` process. Also, assuming that the two shard servers are listening to the `27000` and `27001` ports, connect from the shell to these two processes. So we have a total of three shells opened, one connected to the `mongos` process and two to these individual shards.

We are using the `test` database for this recipe, and sharding has to be enabled on this database. If it's not, then you need to execute the following commands on the shell connected to the `mongos` process:

```
mongos> use test
mongos> sh.enableSharding('test')
```

Chapter 4

How to do it...

1. From the shell connected to the `mongos` process, execute the following two commands:

   `mongos> db.testCol.insert({i : 1})`

   `mongos> sh.status()`

2. In the databases, look out for the `test` database and take note of the primary. Suppose that the following is a part (showing the part under databases only) of the output of `sh.status()`:

   `databases:`

   `{  "_id" : "admin",  "partitioned" : false,  "primary" : "config" }`

   `{  "_id" : "test",  "partitioned" : true,  "primary" : "shard0000" }`

3. The second document under the databases shows us that the `test` database is enabled for sharding (because partitioned is `true`) and the primary shard is `shard0000`.

4. The primary shard, which is `shard0000` in our case, is the `mongod` process listening to port `27000`. Open the shell connected to this process and execute the following query:

   `> db.testCol.find()`

5. Now connect to another `mongod` process listening to port `27001` and execute the following query again:

   `> db.testCol.find()`

6. Note that the data will be found only on the primary shard and not on any other shard.

7. Execute the following command from the Mongos shell:

   `mongos> use admin`

   `mongos> db.runCommand({movePrimary:'test', to:'shard0001'})`

8. Execute the following command again from the Mongo shell connected to the `mongos` process:

   `mongos> sh.status()`

9. From the shell connected to the `mongos` processes running on ports `27000` and `27001`, execute the following query:

   `> db.testCol.find()`

Administration

How it works…

We started a sharded setup and connected to it from the `mongos` process. We started by inserting a document in the `testCol` collection that is not enabled for sharding in the `test` database, which is not enabled for sharding as well. In such cases, the data lies on a shard called the primary shard. Do not mistake this for the primary of a replica set. This is a shard (that itself can be a replica set), and it is the shard chosen by default for all databases and collections for which sharding is not enabled.

When we add the data to a nonsharded collection, it is seen only on the shard that is primary. Executing `sh.status()` tells us the primary shard. To change the primary, we need to execute a command from the `admin` database from the shell connected to the `mongos` process. The command is as follows:

```
db.runCommand({movePrimary:'<database whose primary shard is to be changed>', to:'<target shard>'})
```

Once the primary shard is changed, all existing data in nonsharded databases and collections is migrated to the new primary, and all subsequent writes to nonsharded collections will go to this shard.

Use this command with caution, as it will migrate all the unsharded collections to the new primary, which may take time for big collections.

Manually splitting and migrating chunks

Though MongoDB does a good job by default of splitting and migrating chunks across shards to maintain the balance, under some circumstances, such as a small number of documents or a relatively large number of small documents, where the automatic balancer doesn't split the collection, an administrator might want to split and migrate the chunks manually. In this recipe, we will see how to split and migrate the collection manually across shards. Again, for this recipe, we will set up a simple shard as we saw in *Chapter 1, Installing and Starting the MongoDB Server*.

Getting ready

Refer to the *Starting a simple sharded environment of two shards* recipe in *Chapter 1, Installing and Starting the MongoDB Server*, to set up and start a sharded environment. It is preferred to start a clean environment without any data in it. From the shell, connect to the started `mongos` process.

How to do it...

1. Connect to the `mongos` process from the Mongo shell and enable sharding on the `test` database and the `splitAndMoveTest` collection as follows:

   ```
   > sh.enableSharding('test')
   > sh.shardCollection('test.splitAndMoveTest', {_id:1}, false)
   ```

2. Let us load the data in the collection as follows:

   ```
   > for(i = 1; i <= 10000 ; i++) db.splitAndMoveTest.insert({_id : i})
   ```

3. Once the data is loaded, execute the following command:

   ```
   > db.splitAndMoveTest.find().explain()
   ```

   Note that the number of documents in the two shards in the plan. The value to lookout for in the two documents under the shards key is the result of the explain plan. Within these two documents, the field to look out for is n.

4. Execute the following commands to see the splits of the collection:

   ```
   > config = db.getSisterDB('config')
   > config.chunks.find({ns:'test.splitAndMoveTest'}).pretty()
   ```

5. Split the chunk into two at `5000`, as follows:

   ```
   > sh.splitAt('test.splitAndMoveTest', {_id:5000})
   ```

6. Splitting it doesn't migrate it to the second server. See exactly what happens with the chunks, by executing the following query again:

   ```
   > config.chunks.find({ns:'test.splitAndMoveTest'}).pretty()
   ```

7. We will now move the second chunk to the second shard:

   ```
   > sh.moveChunk('test.splitAndMoveTest', {_id:5001}, 'shard0001')
   ```

8. Execute the following query again and confirm the migration:

   ```
   > config.chunks.find({ns:'test.splitAndMoveTest'}).pretty()
   ```

9. Alternatively, the following explain plan will show a split of about 50-50 percent:

   ```
   > db.splitAndMoveTest.find().explain()
   ```

Administration

How it works...

We simulate a small data load by adding monotonically increasing numbers and discover that the numbers are not split across two shards evenly, by viewing the query plan. This is not a problem, as the chunk size needs to reach a particular threshold, 64 MB by default, before the balancer decides to migrate the chunks across the shards to maintain balance. This is pretty perfect because, in the real world, when the data size gets huge, we will see that eventually, over a period of time, the shards are well balanced.

However, under some circumstances, when the administration decides to split and migrate the chunks; it is possible to do it manually. The two helper functions `sh.splitAt` and `sh.moveChunk` are there to do this work. Let us look at their signatures and see what they do.

The `sh.splitAt` function takes two parameters. The first parameter is the namespace, which has the format `<database>.<collection name>` and the second parameter is the query that acts as the split point to split the chunk into two, possibly two uneven, portions, depending on where the given document is in the chunk. There is another method named `sh.splitFind` that will try and split the chunk in two equal portions.

However, splitting doesn't mean the chunk moves to another shard. It just breaks one big chunk into two, but the data stays on the same shard. It is an inexpensive operation that involves updating the config DB.

The next operation we execute is to migrate the chunk to a different shard after we split it into two. The `sh.MoveChunk` operation is used just to do that. This function takes three parameters. The first one is again the namespace of the collection, which has the format `<database>.<collection name>`; the second parameter is a query a document, whose chunk would be migrated; and the third parameter is the destination chunk.

Once the migration is done, the query's plan shows us that the data is split into two chunks.

Performing domain-driven sharding using tags

The *Starting a simple sharded environment of two shards* and *Connecting to a shard from the Mongo shell and performing operations* recipes in *Chapter 1, Installing and Starting the MongoDB Server*, explained how to start a simple two-server shard and then insert data in a collection after choosing a shard key. The data that gets sharded is more technical, where the data chunk is kept to a manageable size by Mongo, by splitting it into multiple chunks and migrating the chunks across shards to keep the chunk distribution even across shards. However, what if we want the sharding to be more domain-oriented? Suppose we have a database for storing postal addresses and we shard based on postal codes, where we know the postal code range of a city. What we can do is tag the shard servers according to the city name as the tag, add the shard range (postal codes), and associate this range with the tag.

Chapter 4

This way, we can state which servers can contain the postal addresses of which cities. For instance, we know that for Mumbai, being the most populous city, the number of addresses would be huge and thus we add two shards for Mumbai. On the other hand, one shard should be enough to cope with the volumes of Pune; so for now we tag just one shard. In this recipe, we will see how to achieve this use case using tag-aware sharding. If the description is confusing, don't worry; we will see how to implement what we just discussed.

Getting ready

Refer to the *Starting a simple sharded environment of two shards* recipes in *Chapter 1, Installing and Starting the MongoDB Server*, for how to start a simple shard. However, for this recipe, we will add an additional shard. So, we will now start three MongoDB servers listening to ports 27000, 27001, and 27002. Again, it is recommended to start off with a clean database. For the purpose of this recipe, we will be using the userAddress collection to store the data.

How to do it...

1. Assuming that we have three shards up and running; let us execute the following commands:

   ```
   mongos> sh.addShardTag('shard0000', 'Mumbai')
   mongos> sh.addShardTag('shard0001', 'Mumbai')
   mongos> sh.addShardTag('shard0002', 'Pune')
   ```

2. With the tags defined, let us define the range of the pin codes that will map to a tag, as follows:

   ```
   mongos> sh.addTagRange('test.userAddress', {pincode:400001}, {pincode:400999}, 'Mumbai')
   mongos> sh.addTagRange('test.userAddress', {pincode:411001}, {pincode:411999}, 'Pune')
   ```

3. Enable sharding for the test database and userAddress collection as follows:

   ```
   mongos> sh.enableSharding('test')
   mongos> sh.shardCollection('test.userAddress', {pincode:1})
   ```

4. Insert the following documents in the userAddress collection:

   ```
   mongos> db.userAddress.insert({_id:1, name: 'Varad', city: 'Pune', pincode: 411001})
   mongos> db.userAddress.insert({_id:2, name: 'Rajesh', city: 'Mumbai', pincode: 400067})
   mongos> db.userAddress.insert({_id:3, name: 'Ashish', city: 'Mumbai', pincode: 400101})
   ```

Administration

5. Execute the following explain plans:

   ```
   mongos> db.userAddress.find({city:'Pune'}).explain()
   mongos> db.userAddress.find({city:'Mumbai'}).explain()
   ```

How it works...

Suppose we want to partition data driven by domain in a shard; we can use tag-aware sharding. It is an excellent mechanism that lets us tag the shards and then split the data range across shards identified by the tags. We don't really have to bother about the actual machines and their addresses hosting the shard. Tags act as a good abstraction, in the way, we can tag a shard with multiple tags and one tag can be applied to multiple shards.

In our case, we have three shards and we apply tags to each of them using the `sh.addShardTag` method. This method takes the shard ID, which we can see in the `sh.status` call with the "shards" key. This `sh.addShardTag` can be used to keep adding tags to a shard. Similarly, there is a helper method `sh.removeShardTag` to remove an assignment of the tag from the shard. Both these methods take two parameters, the first one is the shard ID and the second one of the tag to remove.

Once the tagging is done, we assign the range of the values of the shard key to the tag. The `sh.addTagRange` method is used to do that. It accepts four parameters; the first one is the namespace (the fully qualified name of the collection), the second and third parameters are the start and end values of the range for this shard key, and the fourth parameter is the tag name of the shards hosting the range being added. For example, the call `sh.addTagRange('test.userAddress', {pincode:400001}, {pincode:400999}, 'Mumbai')` says we are adding the shard range from `400001` to `400999` for the `test.userAddress` collection and this range will be stored in the shards tagged as `Mumbai`.

Once the tagging and adding tag ranges are done, we enable sharding on the database and collection, and add data to it from Mumbai and Pune with the respective pin codes. We then query and explain the plan, to see that the data did indeed reside on the shards we have tagged for Pune and Mumbai city. We can also add new shards to this sharded setup and accordingly tag the new shard. The balancer will then accordingly balance the data based on the value it has tagged. For instance, if the addresses in Pune increase, thus overloading a shard, we can add a new shard with the tag as `Pune`. The postal addresses for Pune will then be sharded across these two server instances tagged for Pune city.

Exploring the config database in a sharded setup

The config database is the backbone of a sharded setup in Mongo. It stores all the metadata of the shard setup and has a dedicated mongod process running for it. When a mongos process is started, we provide it with the config server's URL. In this recipe, we will take a look at some collections in the config database and dive deep into their content and significance.

Getting ready

We will have a sharded setup for this recipe. Refer to the *Starting a simple sharded environment of two shards* in *Chapter 1, Installing and Starting the MongoDB Server*, for how to start a simple shard. Additionally, connect to the mongos process from a shell.

How to do it...

1. From the console connected to the mongos process, switch to the config database and execute view all collections as follows:

   mongos> use config

   mongos>show collections

2. From the list of all collections, we will visit a few. We start with the databases collection. This keeps a track of all the databases on this shard. Execute the following command from the shell:

   mongos> db.databases.find()

   The content of the result is pretty straightforward. The value of the _id field is for the database. The value of the field partitioned tells us whether sharding is enabled for the database or not; true indicates it is enabled and the field primary gives the primary shard where the data of nonsharded collections resides.

3. The next collection we will visit is collections. Execute the following command from the shell:

   mongos> db.collections.find().pretty()

   This collection, unlike the databases collection we saw earlier, contains only those collections for which we have enabled sharding. The _id field gives the namespace of the collection in the <database>.<collection name> format, the key field gives the shard key, and the unique field indicates whether the shard key is unique of not. These three fields come as the three parameters of the sh.shardCollection function in that very order.

Administration

4. Next, we look at the `chunks` collection. Execute the following command on the shell. If the database was clean when we started this recipe, we won't have a lot of data in this:

   `mongos> db.chunks.find().pretty()`

5. We then look at the `tags` collection. Execute the following query:

   `mongos> db.tags.find().pretty()`

6. Let us query the `mongos` collection as follows. This is a simple collection giving the list of all `mongos` instances connected to the shard, with the details such as the hostname and port on which the `mongos` instance is running, which forms the `_id` field, and the version and figures, such as for how much time the process has been up and running in seconds.

   `mongos> db.mongos.find().pretty()`

7. Finally, we look at the `version` collection. Execute the following query (note that it is not similar to other queries we execute):

   `mongos> db.getCollection('version').findOne()`

How it works...

We saw the collections and databases collection when we queried them; they are pretty simple. Let us look at the collection called `chunks`. The following is a sample document from this collection:

```
{
        "_id" : "test.userAddress-pincode_400001.0",
        "lastmod" : Timestamp(1, 3),
        "lastmodEpoch" : ObjectId("53026514c902396300fd4812"),
        "ns" : "test.userAddress",
        "min" : {
                "pincode" : 400001
        },
        "max" : {
                "pincode" : 411001
        },
        "shard" : "shard0000"
}
```

Chapter 4

The fields of interest are `ns` (the namespace of the collection), `min` (the minimum value present in the chunk), `max` (the maximum value present in the chunk), and `shard` (the shard on which this chunk lies). The value of the chunk size is 64 MB by default. This can be seen in the settings collection. Execute `db.settings.find()` from the shell and look at the value of the field value, which is the size of the chunk in MB. Chunks are restricted to this small size to ease the migration process across shards if needed. When the size of the chunk exceeds this threshold, the MongoDB server finds a suitable point in the existing chunk to break it into two, and adds a new entry in this chunk's collection. This operation is called splitting and is inexpensive, as the data stays where it is. It is just logically split into multiple chunks. The balancer on Mongo tries to keep the chunks across shards balanced, and the moment it sees some imbalance, it migrates these chunks to a different shard, which is expensive and also depends largely on the network bandwidth. If we execute `sh.status()`, the implementation actually queries the collections we saw earlier and prints the formatted result.

5
Advanced Operations

In this chapter, we will cover the following recipes:

- Atomic find and modify operations
- Implementing atomic counters in MongoDB
- Implementing server-side scripts
- Creating and tailing capped collection cursors in MongoDB
- Converting a normal collection to a capped collection
- Storing binary data in MongoDB
- Storing large data in Mongo using GridFS
- Storing data to GridFS from a Java client
- Storing data to GridFS from a Python client
- Implementing triggers in MongoDB using oplog
- Executing flat plane (2D) geospatial queries in Mongo using geospatial indexes
- Spherical indexes and GeoJSON-compliant data in MongoDB
- Implementing a full-text search in MongoDB
- Integrating MongoDB with Elasticsearch for a full-text search

Advanced Operations

Introduction

In *Chapter 2, Command-line Operations and Indexes*, we saw how to perform basic operations from the shell to query, update, and insert documents. We also explored different types of indexes and index creation. In this chapter, we go ahead and see some of the advanced features of Mongo, such as GridFS, geospatial indexes, and full-text search. Other recipes we will see include an introduction to capped collections and their uses and implementing server-side scripts in MongoDB.

Atomic find and modify operations

In *Chapter 2, Command-line Operations and Indexes*, we had some recipes that explained various CRUD operations that we perform in MongoDB. There was one concept that we didn't cover that is, atomically finding and modifying documents. Modify consists of both `update` and `delete` operations. In this recipe, we will see `find` and `modify` operations in some detail and, in the next recipe, *Implementing atomic counters in MongoDB*, we will put them to use in implementing counters.

Getting ready

Refer to the *Single node installation of MongoDB* recipe in *Chapter 1, Installing and Starting the MongoDB Server*, and start a single instance of MongoDB. That is the only prerequisite for this recipe. Start a Mongo shell and connect to the started server.

How to do it...

We will test a document in the `atomicOperationsTest` collection as follows:

1. Execute the following commands from the Mongo shell:

    ```
    > db.atomicOperationsTest.drop()
    > db.atomicOperationsTest.insert({i:1})
    ```

2. Execute the following commands from the Mongo shell and observe the output:

    ```
    > db.atomicOperationsTest.findAndModify({
        query: {i: 1},
        update: {$set : {text : 'Test String'}},
        new: false
      }
    )
    ```

3. We will execute another one this time, but with slightly different parameters; observe the output this time around:

   ```
   > db.atomicOperationsTest.findAndModify({
       query: {i: 1},
       update: {$set : {text : 'Updated String'}}, fields: {i: 1, text :1, _id:0},
       new: true
     }
   )
   ```

4. We will execute another update this time that will upsert the document as follows:

   ```
   > db.atomicOperationsTest.findAndModify({
       query: {i: 2},
       update: {$set : {text : 'Test String'}},
       fields: {i: 1, text :1, _id:0},
       upsert: true,
       new: true
     }
   )
   ```

5. Now query the collection once, as follows, and see the documents present:

   ```
   > db.atomicOperationsTest.find().pretty()
   ```

6. We will finally execute the `delete` operation as follows:

   ```
   > db.atomicOperationsTest.findAndModify({
       query: {i: 2},
       remove: true,
       fields: {i: 1, text :1, _id:0},
       new: false
     }
   )
   ```

How it works...

If we perform the `find` and `update` operations independently by first finding the document and then updating it in MongoDB, the results might not be as expected as there might be an interleaving update between the `find` and the `update` operations that will change the document state. In some of the specific use cases, such as implementing atomic counters, this is not acceptable and thus, we need a way to atomically find, update, and return a document. The returned value is either the one before the update is applied or after the update is applied, and this is decided by the invoking client.

Now that we have executed the steps in the preceding section, let us see what we actually did and what all these fields in the JSON document, which are passed as parameters to the `findAndModify` operation, mean. Starting with step 2, we gave a document as a parameter to the `findAndModify` function that contains the following fields. The fields `query`, `update`, and `new` are used to specify the query that will be used to find the document, the update that will be applied to it, and a Boolean value that will be used to specify whether the document returned by the operation is the one after the update is applied or before it was applied. In this case, the value of the new flag is `false`. The resulting document returned is the one before the update is applied.

In step 3, we actually added a new field to the document, passed as a parameter called `fields`, that is used to select a limited set of fields from the resulting document returned. Also, the value of the `new` field is `true`, which indicates that we want the updated document; that is, the one after the `update` operation is executed and not the one before the update.

In step 4, the parameter contained a new field called `upsert`, which upserted (update + insert) the document. That is, if the document with the given query is found, it is updated; otherwise, a new one is created and updated. If the document didn't exist and an upsert happens, having the value of the `new` parameter as `false` will return `null`. This is because there was nothing present before the `update` operation was executed.

Finally, in step 6, the parameter didn't have the `update` field but had the `remove` field with the value `true`, indicating that the document is to be removed. Also, the value of the `new` field was `false`, which means that we expect the document that got deleted.

See also

- The *Implementing atomic counters in MongoDB* recipe, to see how to implement the use case that is used to develop an atomic counter in Mongo

Implementing atomic counters in MongoDB

Atomic counters are a necessity for a large number of use cases. Mongo doesn't have a built-in feature for atomic counters; nevertheless, it can be easily implemented using some of its cool offerings. In fact, implementing it is merely a couple of lines of code. Refer to the previous recipe to know what atomic `find` and `modify` operations are in Mongo.

Getting ready

Refer to the *Single node installation of MongoDB* recipe in *Chapter 1, Installing and Starting the MongoDB Server*, and start a single instance of Mongo. This is the only prerequisite for this recipe. Start a Mongo shell and connect to the started server.

How to do it...

1. Execute the following piece of code from the Mongo shell:

   ```
   > function getNextSequence(counterId) {
       return db.counters.findAndModify(
         {
           query: {_id : counterId},
           update: {$inc : {count : 1}},
           upsert: true,
           fields:{count:1, _id:0},
           new: true
         }
       ).count
     }
   ```

2. Now, from the shell, invoke the following commands:

   ```
   > getNextSequence('Posts Counter')
   > getNextSequence('Posts Counter')
   > getNextSequence('Profile Counter')
   ```

How it works...

The function is as simple as a `findAndModify` operation on a collection used to store all the counters. The counter ID is the `_id` field of the document stored, and the value of the counter is stored in the `count` field. The document passed to `findAndModify` accepts the query that uniquely identifies the document storing the current count, which is a query using the `_id` field. The `update` operation is an `$inc` operation that will increment the value of the `count` field by one, but what if the document doesn't exist? This will happen during the first invocation of the counter. To take care of this scenario, we will be setting the `upsert` flag to `true`, which atomically either updates the document field or creates one. The value, thus, will always start with one, and there are no ways in this function by which we can have any user-defined start number for the sequence and a custom-incremented step. To address such requirements, we will have to specifically add a document with the initialized values to the `counters` collection. Finally, we are interested in the state of the counter after the value is incremented. Hence, we set the value of the `new` field as `true`.

On invoking this method three times, as we did, we should see the following in the `counters` collection. Simply execute the following query:

```
> db.counters.find()
{ "_id" : "Posts Counter", "count" : 2 }
{ "_id" : "Profile Counter", "count" : 1 }
```

Using this small function, we have now implemented atomic counters in Mongo.

We can store such common code on the Mongo server, which will be available for execution in other functions.

See also

- The *Implementing server-side scripts* recipe to see how we can store JavaScript functions on the Mongo server

Implementing server-side scripts

In this recipe, we will see how to write server-stored JavaScript that is similar to stored procedures in relational databases. This is a common use case, where other pieces of code require access to these common functions and we have them in one central place. The function for demo purpose is simple; we will add two numbers. There are two parts to this recipe. First, we'll see how to load the scripts from the collections on the client-side JavaScript shell and then, we will see how to execute these functions on the server.

The documentation specifically mentions that it is not recommended to use server-side scripts. Security is one concern though if the data is not properly audited and, hence, need to be careful about what functions are defined. Since the launch of Mongo 2.4, the server-side JavaScript engine is V8, which can execute multiple threads in parallel, as opposed to the engine prior to Version 2.4 of Mongo, which executes only one thread at a time.

Getting ready

Refer to the *Single node installation of MongoDB* recipe in *Chapter 1, Installing and Starting the MongoDB Server*, and start a single instance of Mongo. This is the only prerequisite for this recipe. Start a Mongo shell and connect to the started server.

How to do it...

1. Create a new function called `add` and save it to the `db.system.js` collection as follows. The current database should be `test`:

    ```
    > use test
    > db.system.js.save({ _id : 'add', value : function(num1, num2) {return num1 + num2}})
    ```

2. Now that this function is defined, load all the functions as follows:

    ```
    > db.loadServerScripts()
    ```

3. Invoke `add` and see if it works:

    ```
    > add(1, 2)
    ```

4. We will use the `add` function and execute this on the server side instead. Execute the following commands from the shell:

   ```
   > use test
   > db.eval('return add(1, 2)')
   ```

5. Execute the following commands:

   ```
   > use test1
   > db.eval('return add(1, 2)')
   ```

How it works...

The `system.js` collection is a plain old collection just like any other collection. We add a new server-side JavaScript using the `save` function in this collection. The `save` function is just a convenience function that inserts the document if it is not present or updates an existing one. The objective is to add a new document to this collection, which you may add using `insert` or `upsert`.

The secret lies in the `loadServerScripts` method. The method executes the following line of code:

```
this.system.js.find().forEach(function(u){eval(u._id + " = " + u.value);});
```

This evaluates JavaScript using the `eval` function, and it assigns the function defined in the `value` attribute of the document to a variable named with the name given in the `_id` field of the document, for each document present in the `system.js` collection.

For example, if the `{ _id : 'add', value : function(num1, num2) {return num1 + num2}}` document is present in the `system.js` collection, the function given in the `value` field of the document will be assigned to the variable named as `add` in the current shell. The `add` value is given in the `_id` field of the document.

These scripts do not really execute on the server, but their definition is stored on the server in a collection. The `loadServerScripts` method just instantiates some variables in the current shell and makes those functions available for invocation. It is the JavaScript interpreter of the shell that executes these functions and not the server. The `system.js` collection is defined in the scope of the database, but once loaded, these are JavaScript functions defined in the shell and hence, the functions are available throughout the scope of the shell, irrespective of the database currently active.

As far as security is concerned, if the shell is connected to the server with security enabled, the user invoking `loadServerScripts` must have privileges to read the collections in the database. For more details on enabling security and various roles a user can have, refer to the *Setting up users in MongoDB* recipe in *Chapter 4, Administration*. As we saw earlier, the `loadServerScripts` function reads data from the `system.js` collection and thus, if the user doesn't have privileges to read from the collection, the function invocation will fail. Apart from that, the functions executed from the shell after being loaded should have appropriate privileges. For instance, if a function inserts/updates in any collection, the user should have read and write privileges on that particular collection accessed from the function.

Executing scripts on the server is perhaps what one would expect to be the server-side script. as against executing in the connected shell. In this case, the functions are evaluated on the server's JavaScript engine and the security checks are more stringent as long-running functions can hold locks having detrimental effects on the performance. The wrapper to invoke the execution of the JavaScript code on the server side is the `db.eval` function, accepting the code to evaluate on the server side along with the parameters, if any.

Before evaluating the function, the write operation takes a global lock. This can be skipped if the `nolock` parameter is used. For instance, the preceding `add` function can be invoked as follows, instead of calling `db.eval`, and will achieve the same results. Additionally, we provided the `nolock` field to instruct the server not to acquire the global lock before evaluating the function. If the function performs any read/write operations on the collection, it will acquire locks as usual, and this field doesn't affect this behavior.

```
> db.runCommand({eval: function (num1, num2) {return num1 + num2}, args:[1, 2],nolock:true})
```

If security is enabled on the server, the invoking user needs to have four roles, namely, `userAdminAnyDatabase`, `dbAdminAnyDatabase`, `readWriteAnyDatabase`, and `clusterAdmin` (on the `admin` database), to successfully invoke the `db.eval` function.

Programming languages do provide a way for the invocation of such server-side scripts as well using the `eval` function. For instance, in Java API, the `com.mongodb.DB` class has the `eval` method to invoke server-side JavaScript code. Such server-side executions are highly useful when we want to avoid unnecessary network traffic for the data and get the result to the clients. However, too much logic on the database server can quickly make things difficult to maintain and can affect the performance of the server badly.

Advanced Operations

Creating and tailing capped collection cursors in MongoDB

Capped collections are fixed-size collections and they act like queues. The documents added to it are added towards the end of the collection, removing the oldest entry in the collection, if the space allocated to the collection becomes full. They provide fast access to the limited-sized collections even without the use of the index. They are naturally sorted by the order of the insertion, and any retrieval needed on them ordered by time can be retrieved using the `$natural` sort order. The following diagram gives a pictorial representation of a capped collection whose size is enough to hold up to three documents of equal size (which is too small for any practical use, but good for illustration purposes). As we see in the diagram, the collection is similar to a circular queue, where the oldest document is replaced by the newly added document, should the collection become full:

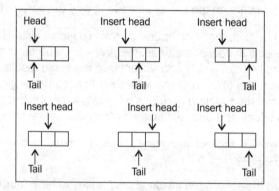

Tailable cursors are a special type of cursor that tails the collection just as a tail command in Unix does. These cursors iterate through the collection like normal cursors do but additionally, they wait for data to be available in the collection if it is not available. We will see capped collections and tailable cursors in detail in this recipe.

Getting ready

Refer to the *Single node installation of MongoDB* recipe in *Chapter 1, Installing and Starting the MongoDB Server*, and start a single instance of Mongo. This is the only prerequisite for this recipe. Start a Mongo shell and connect to the started server.

Chapter 5

How to do it...

There are two parts to this recipe. In the first part, we will be creating a capped collection called `testCapped` and will try performing some basic operations on it. In the second part, we will be creating a tailable cursor on the capped collection we created.

1. First, we will drop it if a collection with the `testCapped` name exists, as follows:

    ```
    > db.testCapped.drop()
    ```

2. Now create a capped collection as follows (note that the size given here is in bytes allocated for the collection and not the number of documents it contains):

    ```
    > db.createCollection('testCapped', {capped : true, size:100})
    ```

3. We will now insert 100 documents in the capped collection, as follows:

    ```
    > for(i = 1; i < 100; i++) {
    db.testCapped.insert({'i':i, val:'Test capped'})
    }
    ```

4. Now query the collection as follows:

    ```
    > db.testCapped.find()
    ```

5. Try to remove the data from the collection, as follows:

    ```
    > db.testCapped.remove()
    ```

    You should get an error after executing the previous command

6. We will now create and demonstrate a tailable cursor. It is recommended that you type/copy the following pieces of code into a text editor and keep it handy for execution.

7. To insert data in a collection, we will be executing the following fragment of code. Execute this piece of code in the shell as follows (note that this execution will take quite some time):

    ```
    > for(i = 101 ; i < 500 ; i++) {
      sleep(1000)
      db.testCapped.insert({'i': i, val :'Test Capped'})
    }
    ```

Advanced Operations

8. To tail a capped collection, we execute the following piece of code:

    ```
    > var cursor = db.testCapped.find().addOption(DBQuery.Option.
    tailable).addOption(DBQuery.Option.awaitData)
    while(cursor.hasNext()) {
      var next = cursor.next()
      print('i: ' + next.i + ', value: ' + next.val)
    }
    ```

9. Open a shell and connect to the running `mongod` process. This will be the second shell opened and connected to the server. Copy and paste the code given in step 8 in this shell and execute it.

10. Observe how the records inserted are shown, as they are inserted into the capped collection.

How it works...

We created a capped collection explicitly using the `createCollection` function. This is the only way a capped collection is created. There are two parameters to the `createCollection` function. The first one is the name of the collection. The second parameter is a JSON document that contains two fields, `capped` and `size`, which are used to inform whether the collection is capped or not, and the size of the collection in bytes, respectively. An additional field `max` can be provided to specify the maximum number of documents in the collection. The field size is required even if the `max` field is specified. We then insert and query the documents. When we try to remove the documents from the collection, we will see an error to the effect that removal is not permitted from the capped collection. It allows the documents to be deleted only when new documents are added, and there isn't space available to accommodate them.

What we see next is about the tailable cursor we created. We started two shells, and one of them is a normal insertion of documents with an interval of 1 second between subsequent insertions. In the second shell, we created a cursor, iterated through it, and printed the documents that we get from the cursor onto the shell. The additional options we added to the cursor make the difference though. There are two options added, `DBQuery.Option.tailable` and `DBQuery.Option.awaitData`. The former option is to instruct that the cursor is tailable rather than normal, where the last position is marked and we can resume where we left off. The latter option waits for more data for some time rather than returning immediately when no data is available and we reach the end of the cursor. The `awaitData` option can be used with tailable cursors only. The combination of these two options gives us a feel similar to the tail command in the Unix filesystem. For a list of different available options, visit http://docs.mongodb.org/manual/reference/method/cursor.addOption/.

See also

- The *Converting a normal collection to a capped collection* recipe

Converting a normal collection to a capped collection

In this recipe, we will demonstrate the process to convert a normal collection to a capped collection.

Getting ready

Refer to the *Single node installation of MongoDB* recipe in *Chapter 1, Installing and Starting the MongoDB Server*, and start a single instance of Mongo. This is the only prerequisite for this recipe. Start a Mongo shell and connect to the started server.

How to do it...

1. Execute the following command to ensure that you are in the `test` database:

   ```
   > use test
   ```

2. Create a normal collection as follows. We will be adding 100 documents to it. Type/copy the following query in the Mongo shell and execute it:

   ```
   for(i = 1 ; i <= 100 ; i++) {
     db.normalCollection.insert({'i': i, val :'Some Text Content'})
   }
   ```

3. Query the collection, as follows, to confirm if it contains the data:

   ```
   > db.normalCollection.find()
   ```

4. Now query the `system.namespaces` collection as follows, and note the result document:

   ```
   > db.system.namespaces.find({name : 'test.normalCollection'})
   ```

5. Execute the following command to convert the collection into a capped collection:

   ```
   > db.runCommand({convertToCapped : 'normalCollection', size : 100})
   ```

6. Query the collection to take a look at the data:

   ```
   > db.normalCollection.find()
   ```

7. Query the `system.namespaces` collection, as follows, and note the result document:

   ```
   > db.system.namespaces.find({name : 'test.normalCollection'})
   ```

Advanced Operations

How it works...

We created a normal collection with 100 documents and then tried to convert it to a capped collection with a size of 100 bytes. The command has the following JSON document passed to the `runCommand` function:

```
{convertToCapped : <name of normal collection>, size: <size in bytes of the capped collection>}
```

This command creates a capped collection with the mentioned size, and loads the documents in the natural order from the normal collection to the target capped collection. If the size of the capped collection reaches the limit mentioned, the old documents are removed in the FIFO order, making space for new documents. Once this is done, the created capped collection is renamed. Executing a `find` query on the capped collection confirms that not all 100 documents that were originally present in the normal collection are present in the capped collection. A query on the `system.namespaces` collection, before and after the execution of the `convertToCapped` command, shows the change in the collection attributes. Note that this operation acquires a global write lock, blocking all read and write operations in this database. Also, any indexes present on the original collection are not created on this capped collection-up conversion.

There's more...

Oplog is an important collection used for replication in MongoDB and is a capped collection. For more information on replication and oplogs, refer to the *Understanding and analyzing oplogs* recipe in *Chapter 4, Administration*. In the *Implementing triggers in MongoDB using oplog* recipe, we will use this oplog to implement a feature similar to the after insert/update/delete trigger of a relational database.

Storing binary data in MongoDB

So far we have seen how to store text values, dates, and numbers fields in a document. Binary content also needs to be stored at times in the database. Consider cases where users upload their photographs or scanned copies of documents that need to be stored in the database. In relational databases, the BLOB data type is the most commonly used type to address these requirements. Mongo too supports binary contents to be stored in a document in the collection. The catch is that the total size of the document shouldn't exceed 16 MB, which is the upper limit of the document size at the time of writing this book. In this recipe, we will store a small image file in Mongo's document and also retrieve it later. If the content you wish to store in MongoDB collections is greater than 16 MB, MongoDB offers an out-of-the-box solution called GridFS. We will see how to use GridFS in the *Storing large data in MongoDB using GridFS* recipe later in this chapter.

Getting ready

Look at the *Single node installation of MongoDB* recipe in *Chapter 1, Installing and Starting the MongoDB Server*, and start a single instance of MongoDB. Also, the program to write binary content to the document is written in Java. For more details on Java drivers, refer to the following recipes in *Chapter 3, Programming Language Drivers*:

- *Executing query and insert operations using a Java client*
- *Executing update and delete operations using a Java client*
- *Aggregation in Mongo using a Java client*
- *MapReduce in Mongo using a Java client*

Open a Mongo shell and connect to the local MongoDB instance listening to port `27017`. For this recipe, we will be using the project `mongo-cookbook-bindata`. This project is available in the source code bundle downloadable from the book's website. The folder needs to be extracted on the local filesystem. Open a command-line shell and go to the root of the project extracted. It should be the directory where the `pom.xml` file is found.

How to do it...

1. On the operating system shell with the `pom.xml` file present in the current directory of the `mongo-cookbook-bindata` project, execute the following command:

    ```
    $ mvn clean compile exec:java -Dexec.mainClass=com.packtpub.mongo.cookbook.BinaryDataTest
    ```

2. Observe the output; the execution should be successful.
3. Switch to the Mongo shell, connected to the local instance, and execute the following query:

    ```
    > db.binaryDataTest.findOne()
    ```

4. Scroll through the document and take a note of the fields in the document.

How it works...

If we scroll through the large document printed out, we see that the fields are `fileName`, `size`, and `data`. The first two fields are of type string and number respectively, which we populated on document creation, and hold the name of the file we provide and the size in bytes. The data field is a field of BSON type, where we see the data encoded in the `base64` format.

Advanced Operations

What we did to insert this document is not much from an application's perspective. The following lines of code show how we populated the DBObject that we added to the collection:

```
DBObject doc = new BasicDBObject("_id", 1);
doc.put("fileName", resourceName);
doc.put("size", imageBytes.length);
doc.put("data", imageBytes);
```

As we see, two fields, namely, `fileName` and `size`, are used to store the name of the file and the size of the file, and are of type string and number respectively. The field data is added to `DBObject` as a byte array. It gets stored automatically as the BSON type `BinData` in the document.

What we saw in this recipe is straightforward, as the document size is less than 16 MB, which is the maximum document size in Mongo as of writing this book. If the size of the files stored exceeds this value, we have to resort to solutions such as GridFS, explained in the next recipe.

Storing large data in MongoDB using GridFS

A document's size in MongoDB can be a maximum of 16 MB, but does that mean we cannot store data that is more than 16 MB in size? There are cases where you prefer to store videos and audio files in a database rather than in the filesystem for a number of advantages, such as, a few of them are storing metadata along with them, accessing the file from an intermediate location, and replicating the contents for high availability if replication is enabled on the MongoDB server instances. GridFS is the way to address such use cases in MongoDB. We will also see how GridFS manages large content that exceeds 16 MB and will analyze the collections it uses for storing the content behind the scenes. For test purposes, we will not be using data exceeding 16 MB but something smaller to see GridFS in action.

Getting ready

Refer to the *Single node installation of MongoDB* recipe in *Chapter 1, Installing and Starting the MongoDB Server*, and start a single instance of Mongo. This is the only prerequisite for this recipe. Start a Mongo shell and connect to the started server. Additionally, we will use the `mongofiles` utility to store data in GridFS from the command line.

How to do it...

1. Download the code bundle of the book from the book's website and save the image file named `glimpse_of_universe-wide.jpg` from it to your local drive (you may choose any other large file, as a matter of fact, and provide an appropriate name to the file with the commands we execute). For the sake of the example, the image is saved in the `Home` directory. We will split our steps into three parts:

2. With the server up and running, execute the following command from the operating system's shell, with the current directory being the Home directory. There are two arguments here. The first one is the name of the file on the local filesystem, and the second one is the name that will be attached to the uploaded content in MongoDB.

   ```
   $ mongofiles put -l glimpse_of_universe-wide.jpg universe.jpg
   ```

3. Let us now query the collections to see how this content is actually stored in the collections behind the scenes. With the shell open, execute the following two queries. Make sure that in the second query, you mention not to select the data field:

   ```
   > db.fs.files.findOne({filename:'universe.jpg'})
   > db.fs.chunks.find({}, {data:0})
   ```

4. Now that we have put a file to GridFS from the operating system's local filesystem, we will see how we can get the file to the local filesystem. Execute the following command from the operating system shell:

   ```
   $ mongofiles get -l UploadedImage.jpg universe.jpg
   ```

5. Finally, we will delete the file we uploaded as follows. From the operating system shell, execute the following command:

   ```
   $ mongofiles delete universe.jpg
   ```

6. Confirm the deletion using the following queries again:

   ```
   > db.fs.files.findOne({filename:'universe.jpg'})
   > db.fs.chunks.find({}, {data:0})
   ```

How it works...

Mongo distribution comes with an out-of-the-box tool called mongofiles that lets us upload large content to the Mongo server; this gets stored using the GridFS specification. GridFS is not a different product but a specification that is standard and followed by different drivers for MongoDB to store data greater than 16 MB, which is the maximum document size. It can even be used for files of size less than 16 MB, as we did in our recipe, but there isn't really a good reason to do that. There is nothing stopping us from implementing our own way of storing these large files, but it is preferred to follow the standard because all drivers support it; they do the heavy lifting of splitting the big file into small chunks and reassembling the chunks when needed.

We kept the image downloaded from the book's website and uploaded it using mongofiles to MongoDB. The command to do that is put, and the -l option gives the name of the file on the local drive that we want to upload. Finally, universe.jpg is the name by which we want the file to be stored on GridFS.

Advanced Operations

On successful execution, we should see the following output:

```
connected to: 127.0.0.1
added file: { _id: ObjectId('5310d531d1e91f93635588fe'), filename: "universe.jpg
", chunkSize: 262144, uploadDate: new Date(1393612082137), md5: "d894ec31b8c5add
d0c02060971ea05ca", length: 2711259 }
done!
```

This gives us some details of the upload, namely, the unique `_id` for the uploaded file, the name of the file, the chunk size (the size of each chunk this big file is broken into, which by default is 256 KB), the date of upload, the checksum of the uploaded content, and the total length of upload. The checksum can be computed beforehand and then compared after the upload, to check whether the uploaded content was corrupted or not.

We executed the following query from the Mongo shell in the `test` database:

```
> db.fs.files.findOne({filename:'universe.jpg'})
```

We see that the output we saw for the `put` command of `mongofiles` is the same as the document queried earlier in the `fs.files` collection. This is the collection where all the uploaded file details are put when some data is added to GridFS. There will be one document per upload. Applications can later also modify this document to add their own custom metadata along with the standard details added by Mongo when adding the data. Applications can very well use this collection to add details. For example, if the document is for an image upload, we can add details such as the name of the photographer, the location where the image was taken, when it was taken, and tags for the individuals in the image in this collection.

The file content is something that contains this data. Let us execute the following query:

```
> db.fs.chunks.find({}, {data:0})
```

We have deliberately left out the `data` field from the result selected. Let us look at the structure of the result document:

```
{
_id: <Unique identifier of type ObjectId representing this chunk>,
file_id: <ObjectId of the document in fs.files for the file whose
chunk this document represent>,
n:<The chunk identifier starts with 0, this is useful for knowing the
order of the chunks>,
data: <BSON binary content for the data uploaded for the file>
}
```

For the file we uploaded, we have 11 chunks of maximum 256 KB each. When a file is being requested, the `fs.chunks` collection is searched by `file_id`, which comes from the `_id` field of the `fs.files` collection, and the `n` field, which is the chunk's sequence. A unique index created on these two fields, when this collection is created for the first time when a file is uploaded using GridFS for fast retrieval of chunks using the file ID, is sorted by the chunk's sequence number.

Similar to `put`, the `get` option is used to retrieve the files from the GridFS and put them on a local filesystem. The option `-l`, which is still used to provide the name, is the name of the file that would be saved on the local filesystem. The final parameter to get the command is the name of the file as stored on GridFS. This is the value of the `filename` field in the `fs.files` collection. Finally, the `delete` command of `mongofiles` simply removes the entry of the file from the `fs.files` and `fs.chunks` collections. The name of the file given for deletion is again the value present in the `filename` field of the `fs.files` collection.

There's more...

Some important use cases of using GridFS are when there are some user-generated contents such as large reports on static data that doesn't change too often and is expensive to generate frequently. Instead of running them all the time, they can be run once and stored until a change in static data is detected, in which case the stored report is deleted and executed again on the next request of the data. The filesystem may not always be available to the application to write the files to, in which case this is a good alternative. There are cases where one might be interested in some intermediate chunk of the data stored, in which case the chunk containing the required data can be accessed. You get some nice features; for instance, the MD5 content of the data is stored automatically and is available for use by the application.

Now that we have seen what GridFS is, let us see some scenarios where using GridFS might not be a very good idea. The performance of accessing the content from MongoDB using GridFS and directly from the filesystem will not be the same. Direct filesystem access will be faster than GridFS, and **proof of concept** (**POC**) for the system to be developed is recommended to measure the performance, see if it is within the acceptable limits; the trade off in performance might be worth the benefits we get. Also, if your application server is fronted with CDN, you might not actually need a lot of I/O for static data stored in GridFS. As GridFS stores the data in multiple documents in collections, atomically updating them is not possible. If we know the content is less than 16 MB, which is the case in a lot of user-generated content or some small files uploaded, we may skip GridFS altogether and store the content in one document as BSON supports the storage of binary content in the document. For more details, refer to the *Storing binary data in MongoDB* recipe.

We will rarely be using the `mongofiles` utility to store, retrieve, and delete data from GridFS. Though it may occasionally be used, the majority of times we will be looking at performing these operations from an application. Thus, in the next couple of recipes, we will see how to connect to GridFS to store, retrieve, and delete files using Java and Python clients.

Advanced Operations

Though this has nothing much to do with Mongo, Openstack is an **Infrastructure as a Service** (**IaaS**) platform and offers a variety of services for computing, storing, networking, and so on. One of the image storage services called Glance supports a lot of persistent stores to store the images. One of the supported stores by Glance is MongoDB's GridFS. You can find more information on how to configure Glance to use GridFS at http://docs.openstack.org/trunk/config-reference/content/ch_configuring-openstack-image-service.html.

See also

- The *Storing data to GridFS from a Java client* recipe
- The *Storing data to GridFS from a Python client* recipe

Storing data to GridFS from a Java client

In the previous recipe, we saw how to store data to GridFS using the command-line utility called mongofiles, which comes with MongoDB to manage large data files. To get an idea of what GridFS is and what collections are used behind the scenes to store the data, refer to the previous recipe. In this recipe, we will look at storing data to GridFS using a Java client. The program will be a highly scaled down version of the mongofiles utility and focuses only on how to store, retrieve, and delete data rather than trying to provide a lot of options such as mongofiles do.

Getting ready

Refer to the *Connecting to a single node from a Java client* recipe from *Chapter 1, Installing and Starting the MongoDB Server,* for all the necessary setup for this recipe. If you are interested in more details on Java drivers, refer to the following recipes in *Chapter 3, Programming Language Drivers*:

- *Executing query and insert operations using a Java client*
- *Executing update and delete operations using a Java client*
- *Aggregation in Mongo using a Java client*
- *MapReduce in Mongo using a Java client*

Open a Mongo shell and connect to the local mongod instance listening to port 27017. For this recipe, we will be using the project mongo-cookbook-gridfs. This project is available in the source code bundle downloadable from the book's website. The folder needs to be extracted on the local filesystem. Open a terminal on your operating system and go to the root of the project extracted. It should be the directory where the pom.xml file is found. Also, save the glimpse_of_universe-wide.jpg file on the local filesystem, just as in the previous recipe. This file can be found in the downloadable code bundle from the book's website.

How to do it...

1. We are assuming that the collections of GridFS are clean and no prior data is uploaded. If there is nothing crucial in the database, you may execute the following queries to clear the collection. Do exercise caution before dropping the collections though:

   ```
   > use test
   > db.fs.chunks.drop()
   > db.fs.files.drop()
   ```

2. Open an operating system shell and execute the following command:

   ```
   $ mvn clean compile exec:java -Dexec.mainClass=com.packtpub.mongo.cookbook.GridFSTests -Dexec.args="put ~/glimpse_of_universe-wide.jpg universe.jpg"
   ```

   The file I need to upload was placed in the Home directory. You may choose to give the filepath of the image file after the put command. Note that if the path contains spaces, the whole path needs to be within single quotes.

3. If the preceding command runs successfully, you should see the following output:

   ```
   Successfully written to universe.jpg, details are:
   Upload Identifier: 5314c05e1c52e2f520201698
   Length: 2711259
   MD5 hash: d894ec31b8c5addd0c02060971ea05ca
   Chunk Side in bytes: 262144
   Total Number Of Chunks: 11
   ```

4. Once the preceding execution is successful, which we can confirm from the console output, execute the following queries from the Mongo shell:

   ```
   > db.fs.files.findOne({filename:'universe.jpg'})
   > db.fs.chunks.find({}, {data:0})
   ```

5. Now we will get the file from GridFS to the local filesystem. Execute the following command to perform this operation. Ensure that the directory to which we are writing, the second parameter after the get operation, is writable to the user.

   ```
   $ mvn clean compile exec:java -Dexec.mainClass=com.packtpub.mongo.cookbook.GridFSTests -Dexec.args="get '~/universe.jpg' universe.jpg"
   ```

Advanced Operations

6. Confirm that the file is present on the local filesystem at the mentioned location. We should see the following output on the console output to indicate a successful write operation:

   `Connected successfully..`

   `Successfully written 2711259 bytes to ~/universe.jpg`

7. Finally, we will delete the file from GridFS as follows:

   `$ mvn clean compile exec:java -Dexec.mainClass=com.packtpub.mongo. cookbook.GridFSTests -Dexec.args="delete universe.jpg"`

8. On successful deletion, we should see the following output on the console:

   `Connected successfully..`

   `Removed file with name 'universe.jpg' from GridFS`

How it works...

The `com.packtpub.mongo.cookbook.GridFSTests` class accepts three types of operations, `put`, `get`, and `delete`, to upload a file to GridFS, get contents from GridFS to a local filesystem, and delete files from GridFS respectively.

The class accepts up to three parameters. The first one is the operation with valid values as `get`, `put`, and `delete`. The second parameter is relevant for the `get` and `put` operations and is the name of the file on the local filesystem to write the downloaded content to, or from which the content is sourced for upload. The third parameter is the name of the file in GridFS, which is not necessarily the same as the name on the local filesystem. However, for `delete`, only the filename on GridFS is needed for deletion purposes.

Let us see some important snippets of code from the class that is specific to GridFS.

Open the `com.packtpub.mongo.cookbook.GridFSTests` class in your favorite IDE and look for the `handlePut`, `handleGet`, and `handleDelete` methods. These are the methods where all the logic is. First, we will start with the `handlePut` method, which is meant to upload the contents of the file from the local filesystem to GridFS.

Irrespective of which operation we do, we will create an instance of the `com.mongodb.gridfs.GridFS` class. In our case, we instantiated it as follows:

```
GridFS gfs = new GridFS(client.getDB("test"));
```

The constructor of this class takes the database instance of the `com.mongodb.DB` class in which we wish to create the GridFS tables `fs.chunks` and `fs.files`, which will store the uploaded content. Once the instance of GridFS is created, we will invoke the `createFile` method on it. This method accepts two parameters; the first one is `InputStream`, which sources the bytes of the content to be uploaded, and the second parameter is the name of the file that will be saved on GridFS. However, this method doesn't create the file on GridFS but returns an instance of `com.mongodb.gridfs.GridFSInputFile`. The upload will happen only when we call the `save` method in this returned object. There are a few overloaded variants of this `createFile` method. For more details, refer to the Javadoc of the `com.mongodb.gridfs.GridFS` class.

Our next method is `handleGet`, which gets the contents of the file saved on GridFS to a local filesystem. Similar to the `com.mongodb.DBCollection` class, the class `com.mongodb.gridfs.GridFS` has the `find` and `findOne` methods to search. However, instead of accepting any `DBObject` query, `find` and `findOne` in GridFS accept the filename or the `ObjectId` value of the document to search in the `fs.files` collection. Similarly, the return value is not a `DBCursor` but an instance of `com.mongodb.gridfs.GridFSDBFile`. This class has various methods that let us get the `InputStream` of the bytes of the file present on GridFS. Other methods are `writeTo`, which writes to the given file or `OutputStream` and a `getLength` method that give the number of bytes in the file. For details, refer to the Javadoc of the `com.mongodb.gridfs.GridFSDBFile` class.

Finally, we look at the `handleDelete` method that is used to delete the files on GridFS and is the simplest of the lot. The method on the object of GridFS is `remove`, which accepts a string argument that is the name of the file to delete on the server. The return type of this method is void. So, irrespective of whether the content is present on GridFS or not, the method will not return a value nor throw an exception if a name is provided to this method for a file that doesn't exist.

See also

- The *Storing binary data in MongoDB* recipe
- The *Storing data to GridFS from a Python client* recipe

Advanced Operations

Storing data to GridFS from a Python client

In the *Storing large data in MongoDB using GridFS* recipe, we saw what GridFS is and how it can be used to store large files in MongoDB. In the previous recipe, we saw how to use GridFS API from a Java client. In this recipe, we will see how to store image data into MongoDB using GridFS from a Python program.

Getting ready

Refer to the *Connecting to a single node from a Java client* recipe from *Chapter 1, Installing and Starting the MongoDB Server*, for all the necessary setup for this recipe. If you are interested in more details on Python drivers, refer to the following recipes in *Chapter 3, Programming Language Drivers*:

- *Installing PyMongo*
- *Executing query and insert operations using PyMongo*
- *Executing update and delete operations using PyMongo*

Download and save the `glimpse_of_universe-wide.jpg` image file from the downloadable code bundle, available on the book's website, to the local filesystem, as we did in the previous recipe.

How to do it...

1. Open a Python interpreter by typing in the following command in the operating system shell (note that the current directory is the same as the directory where the image file `glimpse_of_universe-wide.jpg` is placed):

   ```
   $ python
   ```

2. Import the required packages as follows:

   ```
   >>> import pymongo
   >>> import gridfs
   ```

3. Once the Python shell is opened, create a `MongoClient` and database object to the `test` database as follows:

   ```
   >>> client = pymongo.MongoClient('mongodb://localhost:27017')
   >>> db = client.test
   ```

4. To clear the GridFS-related collections to start clean, and only if nothing important is present in them, execute the following queries:

   ```
   >>> db.fs.files.drop()
   >>> db.fs.chunks.drop()
   ```

Chapter 5

5. Create the instance of GridFS as follows:

    ```
    >>> fs = gridfs.GridFS(db)
    ```

6. Now, we will read the file and upload its contents to GridFS. First, create the `file` object as follows:

    ```
    >>> file = open('glimpse_of_universe-wide.jpg', 'rb')
    ```

7. Now put the file into GridFS as follows:

    ```
    >>> fs.put(file, filename='universe.jpg')
    ```

8. On successfully executing the preceding `put` command, we should see `ObjectId` for the file uploaded. This would be same as the `_id` field of the `fs.files` collection for this file.

9. Execute the following query from the Python shell. It should print out the `dict` object with the details of the upload. Verify the contents and cross-check by executing the following query:

    ```
    >>> db.fs.files.find_one()
    ```

10. Now, we will get the uploaded content and write it to a file on the local filesystem. Let us get the `GridOut` instance representing the object, to read the data out of GridFS as follows:

    ```
    >>> gout = fs.get_last_version('universe.jpg')
    ```

11. With this instance available, let us write the data to the file on a local filesystem as follows. First, open a handle to the file on the local filesystem to write to, as follows:

    ```
    >>> fout = open('universe.jpg', 'wb')
    ```

12. We will then write contents to it as follows:

    ```
    >>> fout.write(gout.read())
    >>> fout.close()
    >>> gout.close()
    ```

13. Now verify the file on the current directory on the local filesystem. A new file called `universe.jpg` will be created with the same number of bytes as the source present in it. Verify it by opening it in an image viewer.

How it works...

Let us look in detail at the steps we executed. In the Python shell, we import two packages, `pymongo` and `gridfs`, and instantiate the `pymongo.MongoClient` and `gridfs.GridFS` instances. The constructor of the `gridfs.GridFS` class takes an argument, which is the instance of `pymongo.Database`.

Advanced Operations

We open a file in binary mode using the `open` function and pass the file object to the GridFS's `put` method. There is an additional argument passed, called `filename`, which is the name of the file put into GridFS. The first parameter, in fact, need not be a `file` object, but any object with a read method defined.

Once the `put` operation succeeds, the return value is an `ObjectId` for the uploaded document in the `fs.files` collection. A query on `fs.files` can confirm that the file is uploaded. Verify that the size of the data uploaded matches the size of the file.

Our next objective is to get the file from GridFS on to the local filesystem. Intuitively, one would imagine that if the method to put a file in GridFS is `put`, then the method to get the file would be `get`. True, the method is indeed `get`. However, it will get only based on the `ObjectId`, which was returned by the `put` method. So if you are ok to get by `ObjectId`, the method for you is `get`. However, if you want to get by the filename, the method to use is `get_last_version`, which accepts the name of the file that we uploaded, and the return type of this method is `gridfs.gridfs_file.GridOut`. This class contains the method read, which will read out all the bytes from the uploaded file to GridFS. We open a file called `universe.jpg` to write in binary mode and write all the bytes read from the `GridOut` object.

See also

- The *Storing binary data in MongoDB* recipe
- The *Storing data to GridFS from a Java client* recipe

Implementing triggers in MongoDB using oplog

A trigger in a relational database is a code that gets invoked when an insert, update, or a delete operation is executed on a table in the database. A trigger can be invoked either before or after the operation. Triggers are not implemented in MongoDB out of the box, and in case you need some sort of notification for your application whenever any `insert`, `update`, and `delete` operations are executed, you are left to manage them by yourself in the application. One approach is to have some sort of data access layer in the application that is the only place to query, insert, update, or delete documents from the collections. However, there are a few challenges to this. First, you need to explicitly code the logic to accommodate this requirement in the application, which may or may not be feasible. If the database is shared and multiple applications access it, things become even more difficult. Second, the access needs to be strictly regulated and no other source of `insert`, `update`, and `delete` should be permitted.

Alternatively, we need to look at running some sort of logic in a layer close to the database. One way to track all write operations is using an oplog. Note that read operations cannot be tracked using oplogs. In this recipe, we will write a small Java application to tail an oplog and get all the `insert`, `update`, and `delete` operations happening on a Mongo instance. Note that this program is implemented in Java and works equally well in any other programming language. The crux lies in the logic for the implementation; the platform for implementation can vary. Also, this works only if the `mongod` instance is started as a part of a replica set and not a standalone instance. Also, this trigger-like functionality can be invoked only after the operation is performed and not before the data gets inserted, updated, or deleted from the collection.

Getting ready

Refer to the *Starting multiple instances as part of a replica set* recipe from *Chapter 1, Installing and Starting the MongoDB Server*, for all the necessary setup for this recipe. If you are interested in more details on Java drivers, refer to the *Executing query and insert operations using a Java client* and *Executing update and delete operations using a Java client* recipes in *Chapter 3, Programming Language Drivers*. The prerequisites of these two recipes are all we need for this recipe.

Refer to the *Creating and tailing capped collection cursors in MongoDB* recipe in this chapter, to know more about capped collections and tailable cursors if you need a refresher. Finally, though not mandatory, *Chapter 4, Administration*, explains oplog in depth in the *Understanding and analyzing oplogs* recipe. This recipe will not explain oplog in depth as we did in *Chapter 4, Administration*. Open a shell and connect it to the primary of the replica set.

For this recipe, we will be using the project `mongo-cookbook-oplogtrigger`. This project is available in the source code bundle downloadable from the book's website. The folder needs to be extracted on the local filesystem. Open a command-line shell and go to the root of the project extracted. It should be in the directory where the `pom.xml` file is found. Also, the `TriggerOperations.js` file will be needed to trigger operations in the database that we intend to capture.

How to do it...

1. Open an operating system shell and execute the following command:

   ```
   mvn clean compile exec:java -Dexec.mainClass=com.packtpub.mongo.cookbook.OplogTrigger -Dexec.args="test.oplogTriggerTest"
   ```

2. With the Java program started, we will open the shell as follows, with the `TriggerOperations.js` file present in the current directory and the `mongod` instance listening to port `27000` as the primary:

   ```
   $ mongo --port 27000 TriggerOperations.js --shell
   ```

Advanced Operations

3. Once the shell is connected, execute the following function we loaded from the JavaScript:

    `test:PRIMARY> triggerOperations()`

4. Observe the output printed out on the console where the Java program `com.packtpub.mongo.cookbook.OplogTrigger` is being executed using Maven.

How it works...

The functionality we implemented is pretty handy for a lot of use cases. Let us see what we did at a higher level first. The Java program `com.packtpub.mongo.cookbook.OplogTrigger` is something that acts as a trigger when new data is inserted, updated, or deleted from a collection in MongoDB. It uses the oplog collection that is the backbone of replication in Mongo to implement this functionality.

The JavaScript we have just acts as a source of producing, updating, and deleting data from the collection. There's nothing really significant to what this JavaScript function does, but it inserts six documents in a collection, updates one of them, deletes one of them, inserts four more documents, and finally, deletes all the documents. You may choose to open the `TriggerOperations.js` file and take a look at how it is implemented. The collection on which it performs is present in the `test` database and is called `oplogTriggerTest`.

When we execute the JavaScript function, we should see something like the following output printed on the console:

```
[INFO] <<< exec-maven-plugin:1.2.1:java (default-cli) @ mongo-cookbook-oplogtriger <<<
[INFO]
[INFO] --- exec-maven-plugin:1.2.1:java (default-cli) @ mongo-cookbook-oplogtriger ---
Connected successfully..
Starting tailing oplog...
Operation is Insert ObjectId is 5321c4c2357845b165d42a5f
Operation is Insert ObjectId is 5321c4c2357845b165d42a60
Operation is Insert ObjectId is 5321c4c2357845b165d42a61
Operation is Insert ObjectId is 5321c4c2357845b165d42a62
Operation is Insert ObjectId is 5321c4c2357845b165d42a63
Operation is Insert ObjectId is 5321c4c2357845b165d42a64
Operation is Update ObjectId is 5321c4c2357845b165d42a60
Operation is Delete ObjectId is 5321c4c2357845b165d42a61
Operation is Insert ObjectId is 5321c4c2357845b165d42a65
Operation is Insert ObjectId is 5321c4c2357845b165d42a66
```

```
Operation is Insert ObjectId is 5321c4c2357845b165d42a67
Operation is Insert ObjectId is 5321c4c2357845b165d42a68
Operation is Delete ObjectId is 5321c4c2357845b165d42a5f
Operation is Delete ObjectId is 5321c4c2357845b165d42a62
Operation is Delete ObjectId is 5321c4c2357845b165d42a63
Operation is Delete ObjectId is 5321c4c2357845b165d42a64
Operation is Delete ObjectId is 5321c4c2357845b165d42a60
Operation is Delete ObjectId is 5321c4c2357845b165d42a65
Operation is Delete ObjectId is 5321c4c2357845b165d42a66
Operation is Delete ObjectId is 5321c4c2357845b165d42a67
Operation is Delete ObjectId is 5321c4c2357845b165d42a68
```

The Maven program runs continuously and never terminates as the Java program doesn't terminate. You may hit *Ctrl + C* to stop the execution.

Let us analyze the Java program, which is where the meat of the content is. The first assumption is that for this program to work, the replica set must be set up, as we will use Mongo's oplog collection. The Java program creates a connection to the primary of the replica set members, connects to the `local` database, and gets the `oplog.rs` collection. Then, all it does is find the last, or nearly the last, timestamp in the oplog. This is done not just to prevent the whole oplog from being replayed on startup, but also to mark a point towards the end in the oplog. The following is the code to find this timestamp value:

```
DBCursor cursor = collection.find().sort(new BasicDBObject("$natural",
-1)).limit(1);
int current = (int) (System.currentTimeMillis() / 1000);
return cursor.hasNext() ? (BSONTimestamp)cursor.next().get("ts") : new
BSONTimestamp(current, 1);
```

The oplog is sorted in the reverse natural order to find the time in the last document in it. As oplogs follow the **FIFO** pattern, sorting the oplog in the natural descending order is equivalent to sorting by the timestamp in descending order.

Once this is done, finding the timestamp as earlier, we query the oplog collection as usual, but with two additional options as follows:

```
DBCursor cursor = collection.find(QueryBuilder.start("ts")
   .greaterThan(lastreadTimestamp).get())
   .addOption(Bytes.QUERYOPTION_TAILABLE)
   .addOption(Bytes.QUERYOPTION_AWAITDATA);
```

Advanced Operations

The query finds all documents greater than a particular timestamp and adds two options `Bytes.QUERYOPTION_TAILABLE` and `Bytes.QUERYOPTION_AWAITDATA`. The latter option can only be added when the former option is added. This not only queries and returns the data, but also waits for some time when the execution reaches the end of the cursor for some more data. Eventually, when no data arrives, it terminates.

During every iteration, store the last seen timestamp as well. This is used when the cursor closes when no more data is available, and we query again to get a new tailable cursor instance. The query this time will use the timestamp that we have stored on the previous iteration when the last document was seen. This process continues indefinitely and we basically tail the collection just as we tail a file in Unix using the `tail` command.

The oplog document contains a field called `op`, for the operation whose values are `i`, `u`, and `d` for insert, update, and delete, respectively. The field `o` contains the inserted or deleted object's ID (`_id`) in the case of insert and delete. In the case the update of the file `o2` contains the `_id`, all we do is simply check for these conditions and print out the operation and the ID of the document inserted, deleted, or updated.

Let's look at things we need to be careful about. Obviously, the deleted documents will not be available in the collection, so `_id` would not really be useful if you intend to query. Also, be careful when selecting a document after the update using the ID we get, as some other operation, later in the oplog, might already have performed more updates on the same document and our application's tailable cursor is yet to reach that point. This is common in case of high-volume systems. We have a similar problem for inserts as well. The document we might query, using the provided id, might be updated/deleted already. Applications using this logic to track these operations must be aware of them.

Alternatively, take a look at the oplog that contains more details, such as the document inserted or the update statement executed. Updates in the oplog collection are idempotent, which means they can be applied any number of times without unintended side effects. For instance, if the actual update was to increment the value by one, the update in the oplog collection will have the set operator with the final value to be expected. This way, the same update can be applied multiple times. The logic you would use would then have to be more sophisticated, to implement such scenarios.

Also, failovers are not handled here. This is needed for production-based systems. The infinite loop on the other hand, opens a new cursor as soon as the first one terminates. There could be a sleep duration introduced before the oplog is queried again, to avoid overwhelming the server with queries. Note that the program given here is not a production-quality code but just a simple demo of the technique that is being used by a lot of other systems to get notified about new data insertions, deletions, and updates to collections in MongoDB.

Chapter 5

MongoDB didn't have the text search feature till version 2.4, and prior to that, all full-text search was handled using external search engines such as Solr or Elasticsearch. Even now, at the time of writing, the text search feature in MongoDB, though production-ready, is something many would still use a dedicated external search indexer. It won't be surprising if a decision is taken to use an external full-text index search tool instead of leveraging MongoDB's inbuilt one. In case of Elasticsearch, the abstraction to flow the data into the indexes is known as river. The MongoDB river in Elasticsearch, which adds data to the indexes as and when the data gets added to the collections in Mongo, is built on the same logic as we saw in the simple program implemented in Java.

Executing flat plane (2D) geospatial queries in Mongo using geospatial indexes

In this recipe, we will see what geospatial queries are and then see how to apply these queries on flat planes. We will put it to use in a test map application.

Geospatial queries can be executed on data in which geospatial indexes are created. There are two types of geospatial indexes. The first one, called 2D indexes, is the simpler of the two. It assumes that the data is given as x,y coordinates. The second one. called 3D or spherical indexes, is relatively more complicated. In this recipe, we will explore 2D indexes and execute some queries on 2D data. The data on which we are going to work is a 25 X 25 grid with some coordinates representing bus stops, restaurants, hospitals, and gardens.

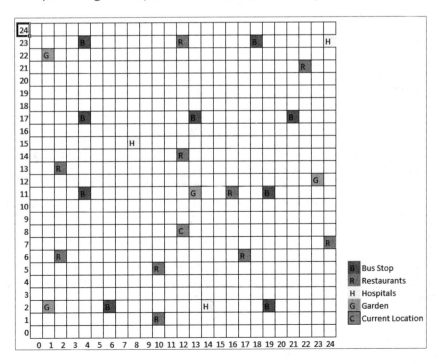

Advanced Operations

Getting ready

Refer to the *Connecting to a single node from a Java client* recipe from *Chapter 1, Installing and Starting the MongoDB Server,* for all the necessary setup for this recipe. Download the data file named 2dMapLegacyData.json and keep it ready to import on the local filesystem. Open a Mongo shell connecting to the local MongoDB instance.

How to do it...

1. Execute the following command from the operating system shell to import the data into the collection. The 2dMapLegacyData.json file is present in the current directory.

   ```
   $ mongoimport -c areaMap -d test --drop 2dMapLegacyData.json
   ```

2. If we see something like the following output on the screen, we can confirm that the import has gone through successfully:

   ```
   connected to: 127.0.0.1
   Mon Mar 17 23:58:27.880 dropping: test.areaMap
   Mon Mar 17 23:58:27.932 check 9 26
   Mon Mar 17 23:58:27.934 imported 26 objects
   ```

3. After the successful import from the opened Mongo shell, verify the collection and its contents by executing the following query:

   ```
   > db.areaMap.find()
   ```

   This should give you the feel of the data in the collection

4. The next step is to create a 2D geospatial index on this data. Execute the following command to create a 2D index:

   ```
   $ db.areaMap.ensureIndex({co:'2d'})
   ```

5. With the index created, we will now try to find the nearest restaurant from the place where an individual is standing. Assuming the person is not fussy about the type of cuisine, let us execute the following query, assuming that the person is standing at location (12, 8), as shown in the preceding screenshot. Also, we are interested in just the three nearest places:

   ```
   $ db.areaMap.find({co:{$near:[12, 8]}, type:'R'}).limit(3)
   ```

6. This should give us three results starting with the nearest restaurant, with the subsequent ones given as per the increasing distance. If we look at the image given earlier, we kind of agree with the results given here.

7. Let us add more options to the query. The individual has to walk and, thus, wants the distance to be restricted to a particular value in the results. Let us rewrite the query with the following modification:

    ```
    $ db.areaMap.find({co:{$near:[12, 8], $maxDistance:4}, type:'R'})
    ```

8. Observe the number of results retrieved this time around.

How it works...

Let us now go through what we did. Before we continue, let us define what exactly we mean by the distance between two points. Suppose on a cartesian plane we have two points, (x1, y1) and (x2, y2), the distance between them would be computed using the following formula:

$$\sqrt{(x_1 - x_2)^2 + (y_1 - y_2)^2}$$

For example, suppose the two points are (2, 10) and (12, 3), the distance would be as follows:

$$\sqrt{(2-12)^2 + (10-3)^2} = \sqrt{(-10)^2 + (7)^2} = \sqrt{149} = 12.207$$

After knowing how calculations for distance are done behind the scenes by MongoDB, let us see what we did right from step 1.

We started by importing the data normally into a `areaMap` collection in the x database and created an index as `db.areaMap.ensureIndex({co:'2d'})`. The index is created on the co field in the document and the value is a special value 2d, which denotes that this is a special type of index called 2D geospatial index. Usually, we give this value as 1, or -1 in other cases, denoting the order of the index.

There are two types of indexes: 2D index and spherical index. A 2D index is commonly used for planes whose span is less and does not involve spherical surfaces. It could be something such as a map of the building, a locality, or even a small city where the curvature of the earth covering the portion of the land is not really significant. However, once the span of the map increases and covers the globe, 2D indexes will be inaccurate for predicting the values, as the curvature of the earth needs to be considered in the calculations. In such cases, we go for spherical indexes, which we will discuss soon.

Once the 2D index is created, we can use it to query the collection and find some points near the point queried, as follows:

```
> db.areaMap.find({co:{$near:[12, 8]}, type:'R'}).limit(3)
```

Advanced Operations

We will query for documents that are of type R, that is, those documents for "restaurants" and that are closest to the point (12, 8). The results returned by this query will be in the increasing order of the distance from the point in question, (12, 8) in this case. The limit just limits the result to the top three documents. We may also provide $maxDistance in the query, which will restrict the results to a distance less than or equal to the provided value. We queried for locations not more than four units away, as follows:

```
> db.areaMap.find({co:{$near:[12, 8], $maxDistance:4}, type:'R'})
```

Spherical indexes and GeoJSON-compliant data in MongoDB

Before we continue with this recipe, we need to look at the previous recipe to get an understanding of what geospatial indexes are in MongoDB and how to use the 2D indexes. What we did so far was to import the JSON documents in a nonstandard format in the MongoDB collection, create geospatial indexes, and query them. This approach works perfectly fine and in fact, was the only option available until MongoDB 2.4. Version 2.4 of MongoDB supports an additional way to store, index, and query the documents in the collections. There is a standard way to represent geospatial data particularly meant for geodata exchange in JSON, and the specification of GeoJSON mentions it in detail at http://geojson.org/geojson-spec.html. We can now store the data in this format.

There are various geographical figure types supported by this specification. However, for our use case, we will be using the type point. First let us see how the document we imported before using a nonstandard format looks and how the one using the GeoJSON format looks:

- Nonstandard way

    ```
    {"_id":1, "name":"White Street", "type":"B", co:[4, 23]}
    ```

- GeoJSON format

    ```
    {"_id":1, "name":"White Street", "type":"B", co:{type: 'Point', coordinates : [4, 23]}}
    ```

The GeoJSON format looks more complicated than the nonstandard format, and for our particular case I do agree. However, when representing polygons and other lines, the nonstandard format might have to store multiple documents; in that case, it can be stored in a single document just by changing the value of the type field. Refer to the specification for more details.

Chapter 5

Getting ready

The prerequisites for this recipe are the same as the prerequisites for the previous recipe, except that the files to be imported will be `2dMapGeoJSONData.json` and `countries.geo.json`. Download these files from the book's website and keep them on the local filesystem to import them later.

> Special thanks to Johan Sundström for sharing the world data. The GeoJSON for the world is taken from `https://github.com/johan/world.geo.json`. The file is massaged to enable importing and index creation in Mongo. Version 2.4 doesn't support `MultiPolygon` and thus, all `MultiPolygon` types of shapes are omitted. This shortcoming seems to be fixed in version 2.6 though.

How to do it...

1. Import the GeoJSON-compatible data in a new collection as follows. This contains 26 documents, similar to what we imported last time around, except that they are formatted using the GeoJSON format.

   ```
   $ mongoimport -c areaMapGeoJSON -d test --drop 2dMapGeoJSONData.json
   $ mongoimport -c worldMap -d test --drop countries.geo.json
   ```

2. Create a geospatial index on this collection as follows:

   ```
   > db.areaMapGeoJSON.ensureIndex({"co" : "2dsphere"})
   > db.worldMap.ensureIndex({geometry:'2dsphere'})
   ```

3. We will first query the `areaMapGeoJSON` collection as follows:

   ```
   > db.areaMapGeoJSON.find(
   {   co:{
         $near:{
           $geometry:{
             type:'Point',
             coordinates:[12, 8]
           }
         }
       },
       type:'R'
   }).limit(3)
   ```

Advanced Operations

4. Next, we will try and find all the restaurants that fall within the square drawn between the points (0, 0), (0, 11), (11, 11), and (11, 0). Refer to the previous screenshot to get a clear visual of the points and the results to expect.

5. Write the following query and observe the results:

```
> db.areaMapGeoJSON.find(
{  co:{
    $geoIntersects:{
      $geometry:{
        type:'Polygon',
        coordinates:[[[0, 0], [0, 11], [11, 11], [11, 0], [0, 0]]]
      }
    }
  },
  type:'R'
})
```

6. Check whether it contains the three restaurants at coordinates (2, 6), (10, 5), and (10, 1) as expected.

7. Next, we will try and perform some operations that would find all the matching objects that lie completely within another enclosing polygon. Suppose we want to find some bus stops that lie within a given square block; such use cases can be addressed using the `$geoWithin` operator, and the query to achieve it is as follows:

```
> db.areaMapGeoJSON.find(
  {co:{
    $geoWithin:{
      $geometry:{
        type: 'Polygon',
        coordinates : [[ [3, 9], [3, 24], [6, 24], [6, 9], [3, 9]]
  ]}
    }
  },
  type:'B'
}
)
```

8. Verify the results; we should have three bus stops in the result. Refer to the preceding screenshot to get the expected results of the query.

Chapter 5

9. When we execute the preceding commands, they just print the documents in the ascending order of the distance. However, we don't see the actual distance in the result. Let us execute the same query as in step 3 and additionally get the calculated distances as follows:

   ```
   > db.runCommand({
       geoNear: "areaMapGeoJSON",
       near: [ 12, 8 ],
       spherical: true,
       limit:3,
       query:{type:'R'}
     }
   )
   ```

10. The preceding query returns one document with an array within the field called results, which contains the matching documents and the calculated distances. The result also contains some additional stats giving the maximum distance, the average of the distances in the result, total documents scanned, and the time taken in milliseconds.

11. We will finally query on the `worldMap` collection to find which country the provided coordinate lies in. Execute the following query from the Mongo shell:

    ```
    db.worldMap.find(
      {geometry:{
        $geoIntersects:{
          $geometry:{
            type:'Point',
            coordinates:[7, 52]
          }
        }
      }
    }
    ,{properties:1, _id:0}
    )
    ```

12. The possible operations we can perform with the `worldMap` collection are numerous, and it is not practically possible to cover all of them in this recipe. I would encourage you to play around with this collection and try out different use cases.

Advanced Operations

How it works...

Starting from MongoDB Version 2.4, the standard way for storing geospatial data in JSON is also supported. Note that the legacy approach we saw is also supported. However, if you are starting fresh, it is recommended that you go ahead with this approach, for the following reasons:

- It is a standard and anybody aware of the specification will easily be able to understand the structure of the document
- It makes storing complex shapes, polygons, and multi lines easy
- It also lets us query easily for the intersection of the shapes, using `$geoIntersect` and other new sets of operators

For using GeoJSON-compatible documents, we import JSON documents in the `2dMapGeoJSONData.json` file into the `areaMapGeoJSON` collection and create the index as follows:

```
> db.areaMapGeoJSON.ensureIndex({"co" : "2dsphere"})
```

The collection has data similar to what we had imported into the `areaMap` collection in the previous recipe, but with a different structure that is compatible with the JSON format. The type used here is 2D sphere and not 2D. The `2dsphere` type of index also considers the spherical surfaces in calculations. Note that the field `co`, on which we are creating the geospatial index, is not an array of coordinates but a document itself that is GeoJSON-compatible.

We query where the value of the `$near` operator is not an array of the coordinates, as we did in the previous recipe, but a document with the `$geometry` key and the value is a GeoJSON-compatible document for a point with the coordinates. The results, irrespective of the query we use, are identical. Refer to step 3 in this recipe and step 5 in the previous recipe, to see the difference in the query. The approach using GeoJSON looks more complicated but it has some advantages, which we will see soon.

It is important to note that we cannot mix the two approaches. Try executing the query in the GeoJSON format we just executed on the `areaMap` collection and note that though we do not get any error, the results are not correct.

We used the `$geoIntersects` operator in step 5 of this recipe. This is only possible when the documents are stored in the GeoJSON format in the database. The query simply finds all the points in our case that intersect any shape we create. We create a polygon using the GeoJSON format, as follows:

```
{
  type:'Polygon',
  coordinates:[[[0, 0], [0, 11], [11, 11], [11, 0], [0, 0]]]
}
```

The coordinates are for the square giving the four corners in a clockwise direction, with the last coordinate the same as the first, denoting it to be complete. The query executed is the same as $near, apart from the fact that the $near operator is replaced by $geoIntersects, and the value of the $geometry field is the GeoJSON document of the polygon with which we wish to find the intersecting points in the areaMapGeoJSON collection. If we look at the results obtained and look at the preceding screenshot, they indeed are what we expected.

The $geoWithin operator (http://docs.mongodb.org/manual/reference/operator/query/geoWithin/) is pretty handy to use when we want to find the points in the polygon or even within another polygon. Note that only shapes completely inside the given polygon will be returned. Suppose that, just like our worldMap collection, we have a cities collection with their coordinates specified in a similar manner. We can then use the polygon of a country to query all the polygons in the cities collection that lie entirely within it, thus giving us the cities. Obviously, an easier and faster way would be store the country code in the city document. Alternatively, if we have some data missing in the cities collection and the country is not present, one point anywhere within the city's polygon (as a city entirely lies in one country) can be used and a query can be executed on the worldMap collection to get its country, which we have demonstrated in step 11.

A combination of what we saw earlier can be put to good use to compute the distances between two points or even execute geometric operation.

Some of the functionalities, such as getting the centroid of a polygon figure, or even the area of a polygon, stored as GeoJSON in the collection, are not supported out of the box and there should have been some utility functions to help compute these given coordinates. These features are good and it is commonly required to have them; perhaps, we might have some support in future releases for operations that can be implemented by developers themselves. Also, there is no straightforward way to find out, if there is an overlap between two polygons, at which coordinates they overlap, what is the area of overlap, and so on. The $geoIntersects operator we saw does tell us which polygons do intersect with the given polygon, point, or line.

Though unrelated to Mongo, the GeoJSON format doesn't have support for circles; hence, storing circles in Mongo using GeoJSON format is not possible. For more details on geospatial operators, visit http://docs.mongodb.org/manual/reference/operator/query-geospatial/.

Advanced Operations

Implementing a full-text search in MongoDB

Many of us (I won't be wrong if I say all of us) use Google every day to search content on the Web. To cut a long story short, the text that we provide in the textbox on Google's page is used to search the pages on the Web that it has indexed. The search results are then returned to us in an order determined by Google's page rank algorithm. We might want to have a similar functionality in our database that lets us search for some text content and gives the corresponding search results. Note that this text search is not the same as finding the text as part of a sentence, which can easily be done using regex. It goes way beyond that and can be used to get results that contain the same, are similar sounding, or have a similar base word; we can even return even a synonym in the actual sentence.

Since MongoDB Version 2.4, the text indexes introduced let us create text indexes on a particular field in the document and enable text search on those words. In this recipe, we will be importing some documents and creating text indexes on them, which we later query to retrieve the results.

Getting ready

A simple single node is what we need for the test. Refer to the *Single node installation of MongoDB* recipe in *Chapter 1, Installing and Starting the MongoDB Server*, for how to start the server. However, do not start the server yet. There is an additional flag provided during the startup to enable text search. Download the `BlogEntries.json` file from the book's website and keep it on your local drive, ready to be imported.

How to do it...

1. Start the MongoDB server listening to port `27017` as follows:

   ```
   $ mongod /data/mongo/db --smallfiles –oplogSize 50 --setParameter textSearchEnabled=true
   ```

   As version 2.4 is used, we need to explicitly enable text search using `textSearchEnabled`. For version 2.6 and above, this command-line option can be skipped.

2. Once the server is started, we will be creating the `test` data in a collection as follows. With the `BlogEntries.json` file placed in the current directory, we will be creating a `userBlog` collection using `mongoimport`:

   ```
   $ mongoimport -d test -c userBlog BlogEntries.json --drop
   ```

3. Now connect to the Mongo process from the Mongo shell, by typing the following command from the operating system shell:

   ```
   $ mongo
   ```

4. Once connected, get a feel of the documents in the `userBlog` collection as follows:

   ```
   > db.userBlog.findOne()
   ```

5. The `blog_text` field is of interest, and this is the one on with which we will be creating a text search index.

6. Create a text index on the `blog_text` field of the document as follows:

   ```
   > db.userBlog.ensureIndex({'blog_text':'text'})
   ```

7. Now execute the following search on the collection from the Mongo shell. The following way is the only way to perform a text search in version 2.4. In version 2.6, though it works fine, it is deprecated.

   ```
   $ db.runCommand({'text':'userBlog', search:'plot zoo'})
   ```

8. Look at the results obtained.

9. Execute another search as follows:

   ```
   $ db.runCommand({'text':'userBlog', search:"Zoo -plot"})
   ```

How it works...

Let us now see how it all works. Text search is done by a process called reverse indexes. In simple terms, this is a mechanism where the sentences are broken into words and then, these individual words point back to the document that they belong to. The process is not straightforward, though; let us see what happens in this process step by step at a high level:

1. Consider the sentence, "I played cricket yesterday". The first step is to break this sentence into tokens as ["I", "played", "cricket", "yesterday"].

2. Next, the stop words from the broken-down sentence are removed, and we are left with a subset of these. Stop words are a list of very common words that are eliminated, as it makes no sense to index them since they can potentially affect the accuracy of the search when used in the search query. In this case, we will be left with the words ["played", "cricket", "yesterday"]. Stop words are language-specific and will be different for different languages.

3. Finally, these words are stemmed to their base words. In this case, it will be ["play", "cricket", "yesterday"]. Stemming is the process of reducing a word to its root. For instance, the words play, playing, played, and plays have the same root wordplay. There are a lot of algorithms and frameworks present for stemming a word to its root form. For more information on stemming and the algorithms used for this purpose, visit `http://en.wikipedia.org/wiki/Stemming`. Similar to eliminating the stop words, the stemming algorithm is language-dependent. The examples given here were for the English language.

If we look at the index creation process, it is as follows:

```
> db.userBlog.ensureIndex({'blog_text':'text'})
```

Advanced Operations

The key given in the JSON argument is the name of the field on which the text index is to be created, and the value will always be text denoting that the index to be created is a text index. Once the index is created, at a high level, the preceding three steps get executed on the content of the field on which the index is created in each document, and a reverse index is created. You may also choose to create a text index on more than one field. Suppose we had two fields, `blog_text1` and `blog_text2`; we can create the index as `{'blog_text1':'text', 'blog_text2':'text'}`. The value `{'$**':'text'}` creates an index on all fields of the document.

Finally, we executed the search operation by invoking the following command:

```
db.runCommand({'text':'userBlog', search:'plot zoo'})
```

The preceding command runs the text search on the `userBlog` collection and the search string used is `plot zoo`. This searches for the value `plot` or `zoo` in the text, in any order. If we look at the results, we see that we have two matched documents and these documents are ordered by the score. This score tells us how relevant the document searched is; the higher the score, the more its relevance. In our case, one of the documents had both the words `plot` and `zoo` in it and thus got a higher score than another document, as we see in the following example. The result also contains a short summary of the total number of documents scanned to get the results, the number of results, and the total time taken to search.

```
{
    "queryDebugString" : "bought|zoo||||||",
    "language" : "english",
    "results" : [
        {
            "score" : 2.6353665865384617,
            "obj" : {
                "_id" : 5,
                ...
            }
        },
        {
            "score" : 0.5263157894736842,
            "obj" : {
                "_id" : 6,
                ...
            }
        }
    ],
```

```
    "stats" : {
        "nscanned" : 3,
        "nscannedObjects" : 0,
        "n" : 2,
        "nfound" : 2,
        "timeMicros" : 119
    },
    "ok" : 1
}
```

In version 2.6, the recommended way to query for the same result is as follows:

```
> db.userBlog.find({$text:{$search:'plot zoo'}})
```

Note that if we compare the result of ordering with the previous execution using `runCommand`, we see that here the results are given in the ascending order of the score. Also, the score is not available in the result that was available in the previous run using `runCommand`. To get the scores in the result, we need to modify the query a bit, as follows:

```
> db.userBlog.find({$text:{$search:'plot zoo'}}, {score: { $meta:
"textScore"}})
```

Now we have an additional document provided in the `find` method that asks for the score calculated for the text match. The results are still not ordered in the descending order of the score. Let us see how to sort the results by score:

```
> db.userBlog.find({$text:{$search:'plot zoo'}}, { score: { $meta:
"textScore" }}).sort({score: { $meta: "textScore"}})
```

As we can see, the query is the same as before. It's just the additional `sort` function we added that will sort the results by the descending order of the score.

When the search was executed as `{'text':'userBlog', search:"Zoo -plot"}`, it searched for all the documents that contain the word `Zoo` but don't contain the word `plot`. Thus we get only one result. The - sign is for negation and leaves out the document from the search result containing that word. However, do not expect to find all documents without the word `plot` by just giving `-plot` in the search.

If we look at the contents returned as a result of the search, they contain the matched documents in entirety. If we are not interested in full documents, but only a few sections of a document, we can use projection to get the desired fields of the document. For instance, use the following query:

```
> db.runCommand({'text':'userBlog', search:"zoo plot", project:{_id:1}})
```

This will be the same as finding all the documents in the `userBlog` collection containing the words `Zoo` or `plot`, but the results will contain the `_id` field from the resulting documents.

Advanced Operations

If multiple fields are used to create an index, then we may have different weights for different fields in the document. For instance, if `blog_text1` and `blog_text2` are two fields, we create an index; we want `blog_text1` given a higher weight than `blog_text 2`, so we create the index as follows:

```
> db.collection.ensureIndex(
  {
    blog_text1: "text",
    blog_text2: "text"
  },
  {
    weights: {
      blog_text1: 2,
      blog_text2: 1,
    },
    name: "MyCustomIndexName"
  }
)
```

This gives the content in `blog_text1` twice as much weight as that in `blog_text2`. Thus, if a word is found in two documents, but is present in the `blog_text1` field of the first document and `blog_text2` of the second document, then the score of first document will be more than that of the second. Note that we have also provided the name of the index using the name field as `MyCustomIndexName`.

We also see from the language key that the language in this case is English. MongoDB supports various languages to implement text search. Languages are important when indexing the content, as they decide the stop words; the stemming of words is language-specific as well. Visit `http://docs.mongodb.org/manual/reference/command/text/#text-search-languages` for more details on the languages supported by Mongo for text search. So how do we choose the language while creating the index? By default, if nothing is provided, the index is created assuming that the language is English. However, if we know the language is French, we create the index as follows:

```
> db.userBlog.ensureIndex({'text':'text'}, {'default_language':'french'})
```

Suppose we had originally created the index using the French language, the `getIndexes` method will return the following document:

```
[
  {
    "v" : 1,
    "key" : {
      "_id" : 1
```

```
      },
      "ns" : "test.userBlog",
      "name" : "_id_"
    },
    {
      "v" : 1,
      "key" : {
        "_fts" : "text",
        "_ftsx" : 1
      },
      "ns" : "test.userBlog",
      "name" : "text_text",
      "default_language" : "french",
      "weights" : {
        "text" : 1
      },
      "language_override" : "language",
      "textIndexVersion" : 1
    }
]
```

However, if the language was different on a per-document basis, which is pretty common in scenarios such as blogs, we have a way out. If we look at the preceding document, the value of the `language_override` field is language. This means that, on a per-document basis, we can store the language of the content using this field. In its absence, the value will be assumed as the default value; French in the preceding case. Thus, we can have:

```
{_id:1, language:'english', text: ….}   //Language is English
{_id:2, language:'german', text: ….}    //Language is German
{_id:3, text: ….}   //Language is the default one; French in this case
```

There's more...

To use MongoDB text search in production, you would need version 2.6. Till version 2.4, the MongoDB text search was in beta. Integrating MongoDB with other systems such as Solr and Elasticsearch is a wise choice to make for now, at least till the text search feature in Mongo matures. In the next recipe, we will see how to integrate Mongo with Elasticsearch, using the Mongo connector.

See also

- For more information on the `$text` operator, visit http://docs.mongodb.org/manual/reference/operator/query/text/

Advanced Operations

Integrating MongoDB with Elasticsearch for a full-text search

MongoDB has integrated text search features, as we saw in the previous recipe. However, there are multiple reasons why one would not use the Mongo text search feature and would fall back to conventional search engines such as Solr or Elasticsearch. The following are a few of the reasons:

- The text search feature is production-ready in version 2.6. In version 2.4, it was introduced in beta, which is not suitable for production use cases.
- Products such as Solr and Elasticsearch are built on top of Lucene, which has proven itself in the search engine arena. Solr and Elasticsearch are pretty stable products too.
- You might already have expertise on products such as Solr and Elasticsearch and would like to use them as full-text search engines rather than MongoDB.
- Some particular feature that your application might require may be missing in MongoDB search,.

Setting up a dedicated search engine does need additional efforts to integrate it with a MongoDB instance. In this recipe, we will see how to integrate a MongoDB instance with the search engine Elasticsearch.

We will be using the Mongo connector for integration purpose. It is an open source project that is available at `https://github.com/10gen-labs/mongo-connector`.

Getting ready

Refer to the *Installing PyMongo* recipe in *Chapter 3, Programming Language Drivers*, to install and set up Python. The tool `pip` is used to get the Mongo connector. However, if you are working on the Windows platform, the steps to install `pip` were not mentioned earlier. Visit `https://sites.google.com/site/pydatalog/python/pip-for-windows` to get `pip` for Windows.

The prerequisites for starting a single instance are all we need for this recipe. However, in this recipe, we will start the server as a single node replica set for demonstration purpose.

Download the `BlogEntries.json` file from the book's website and keep it on your local drive, ready to be imported.

Download Elasticsearch for your target platform from `http://www.elasticsearch.org/overview/elkdownloads/`. Extract the downloaded archive, and from the shell, go to the `bin` directory of the extraction.

Chapter 5

We will be getting the `mongo-connector` source from `github.com` and running it. A Git client is needed for this purpose. Download and install the Git client on your machine. Visit `http://git-scm.com/downloads` and follow the instructions to install Git on your target operating system. If you are not comfortable installing Git on your operating system, then there is an alternative available that lets you download the source as an archive.

Visit `https://github.com/10gen-labs/mongo-connector`. Here, you will get an option that lets you download the current source as an archive, which we can then extract on our local drive. The following screenshot shows the download option available on the bottom-right corner of the screen:

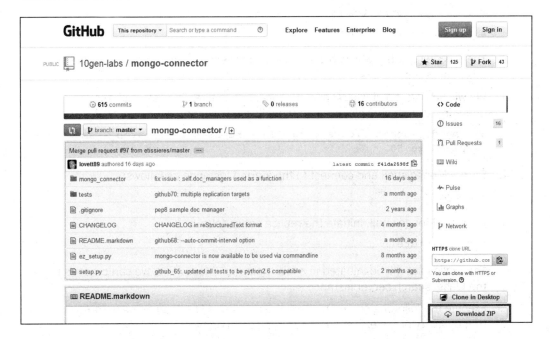

Note that we can also install `mongo-connector` in a very easy way using `pip` as follows:

`pip install mongo-connector`

However, the version in PyPI is very old, with not many features supported and thus, using the latest version from the repository is recommended.

Just like in the previous recipe, where we saw text search in Mongo, we will use the five documents to test our simple search. Download and keep `BlogEntries.json`

Advanced Operations

How to do it...

1. At this point, it is assumed that Python, PyMongo, and `pip` for your operating system platform are installed. We will now get `mongo-connector` from the source. If you have already installed the Git client, we will be executing the following steps on the operating system shell. If you have decided to download the repository as an archive, you may skip this step. Go to the directory where you would like to clone the connector repository, and execute the following commands:

   ```
   $ git clone https://github.com/10gen-labs/mongo-connector.git
   $ cd mongo-connector
   $ python setup.py install
   ```

2. The preceding setup will also install the Elasticsearch client that will be used by this application.

3. We will now start a single Mongo instance, but as a replica set. From the operating system console, execute the following command:

   ```
   $ mongod --dbpath /data/mongo/db --replSet textSearch --smallfiles --oplogSize 50
   ```

4. Start a Mongo shell and connect to the started instance as follows:

   ```
   $ mongo
   ```

5. From the Mongo shell, initiate the replica set as follows:

   ```
   > rs.initiate()
   ```

6. The replica set will be initiated in a few moments. Meanwhile, we can proceed to start the Elasticsearch server instance.

7. Execute the following command from the command line after going to the `bin` directory of the extracted `elasticsearch` archive:

   ```
   $ elasticsearch
   ```

8. We won't be getting into Elasticsearch settings and will start it in the default mode.

9. Once started, enter `http://localhost:9200/_nodes/process?pretty` in the browser.

10. If we see a JSON document, such as the following, giving the process details, we have successfully started Elasticsearch:

```
{
  "cluster_name" : "elasticsearch",
  "nodes" : {
    "p0gMLKzsT7CjwoPdrl-unA" : {
      "name" : "Zaladane",
      "transport_address" : "inet[/192.168.2.3:9300]",
      "host" : "Amol-PC",
      "ip" : "192.168.2.3",
      "version" : "1.0.1",
      "build" : "5c03844",
      "http_address" : "inet[/192.168.2.3:9200]",
      "process" : {
        "refresh_interval" : 1000,
        "id" : 5628,
        "max_file_descriptors" : -1,
        "mlockall" : false
      }
    }
  }
}
```

11. Once the Elasticsearch server and Mongo instance are up and running, and the necessary Python libraries installed, we will start the connector that will sync the data between the started Mongo instance and the Elasticsearch server.

    For the sake of this test, we will be using the `user_blog` collection in the `test` database. The field on which we would like to have text search implemented is the `blog_text` field in the document.

12. Start the Mongo connector from the operating system shell as follows. The following command was executed with the Mongo connector's directory as the current directory:

    ```
    $ python mongo_connector/connector.py -m localhost:27017 -t
    http://localhost:9200 -n test.user_blog --fields blog_text -d
    mongo_connector/doc_managers/elastic_doc_manager.py
    ```

Advanced Operations

13. Import the `BlogEntries.json` file into the collection using the `mongoimport` utility as follows. The command is executed with the `.json` file present in the current directory:

    ```
    $ mongoimport -d test -c user_blog BlogEntries.json --drop
    ```

14. Open a browser of your choice and enter `http://localhost:9200/_search?q=blog_text:facebook` in it.

15. You should see something like the following screenshot in the browser:

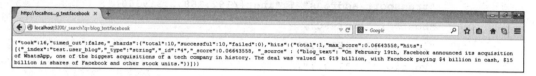

How it works...

Basically, Mongo connector tails the oplog to find new updates that it publishes to another endpoint. We used Elasticsearch in our case, but it could even be Solr. You may choose to write a custom `DocManager` that would plugin with the connector. For more details, visit `https://github.com/10gen-labs/mongo-connector/wiki`. The Readme for `https://github.com/10gen-labs/mongo-connector` gives some detailed information as well.

We gave the connector the `-m`, `-t`, `-n`, `--fields`, and `-d` options. Their meaning as follows:

Option	Description
-m	The URL of the MongoDB host to which the connector connects to get the data to be synchronized.
-t	The target URL of the system with which the data is to be synchronized; Elasticsearch in this case. The URL format will depend on the target system. Should you choose to implement your own `DocManager`, the format will be one that your `DocManager` understands.
-n	This is the namespace that we would like to keep synchronized with the external system. The connector will just be looking for changes in these namespaces while tailing the oplog for data. The value will be separated by commas if more than one namespace is to be synchronized.

Option	Description
`--fields`	These are the fields from the document that would be sent to the external system. In our case, it doesn't make sense to index the entire document and waste resources. It is recommended to add to the index just to the fields where you would like to add text search support. The identifier `_id` field and the namespace of the source are also present in the result, as we can see in the preceding screenshot. The `_id` field can then be used to query the target collection.
`-d`	This is the document manager to be used; in our case, we have used the Elasticsearch's document manager.

For more supported options, refer to the readme of the connector's page on GitHub.

Once the insert is executed on the MongoDB server, the connector detects the newly added documents to the collection of its interest, that is, `user_blog`, and starts sending the data to be indexed from the newly added documents to Elasticsearch. To confirm the addition, we execute a query in the browser to view the results.

Elasticsearch will complain about index names with upper case characters in them. The mongo connector doesn't take care of this and thus, if the name of the collection has to be in lower case (for example, `userBlog`), it will fail.

There's more...

We have not done any additional configuration on Elasticsearch, as that was not the objective of this recipe. We were more interested in integrating MongoDB and Elasticsearch. You will have to refer to the Elasticsearch documentation for more advanced config options. If integration with Elasticsearch is required, there is a concept called rivers in Elasticsearch, that can be used as well. Rivers are Elasticsearch's way to get data from another data source. For MongoDB, the code for a river can be found at `https://github.com/richardwilly98/elasticsearch-river-mongodb/`. `README.md` in this repository has steps on how to set up.

In this chapter, we explored a recipe named *Implementing triggers in Mongo using oplog*, on how to implement trigger-like functionalities using Mongo. This connector and the MongoDB river for Elasticsearch rely on the same logic to get the data out of Mongo as and how it is needed.

See also

- The Elasticsearch documentation at `http://www.elasticsearch.org/guide/en/elasticsearch/reference/`

6
Monitoring and Backups

In this chapter, we will be taking a look at the following recipes:

- Signing up for MMS and setting up the MMS monitoring agent
- Managing users and groups on the MMS console
- Monitoring MongoDB instances on MMS
- Setting up monitoring alerts on MMS
- Backing up and restoring data in Mongo using out-of-the-box tools
- Configuring the MMS backup service
- Managing backups in the MMS backup service

Introduction

Monitoring and backups are important aspects of any mission-critical software in production. Monitoring proactively lets us take actions whenever any abnormal event occurs in the system, which can compromise the data consistency, availability, or the performance of the system. Issues might come to light after having a significant impact in the absence of proactive monitoring of the systems. We covered administration-related recipes in *Chapter 4, Administration*, and both monitoring and backup activities are part of it. However, they demand a separate chapter, as the content to be covered is extensive. In this chapter, we will see how to monitor various parameters and set up alerts for various parameters of your MongoDB cluster, using the **MongoDB Monitoring Service** (**MMS**). We will look at some mechanisms to back up the data using the out-of-the-box tools provided and also using the MMS backup service.

Signing up for MMS and setting up the MMS monitoring agent

MMS is a cloud-based or on-premises service that enables you to monitor your MongoDB cluster. The on-premise version is available with the Enterprise subscription only. It gives you one central place that lets the administrators monitor the health of the server instances and the boxes on which the instances are running. In this recipe, we will see what the software requirements are and how to set up MMS for Mongo.

Getting ready

We will be starting a single instance of mongod, which we will be using for the purpose of monitoring. Refer to the *Single node installation of MongoDB* recipe in *Chapter 1, Installing and Starting the MongoDB Server*, to start a MongoDB instance and connect to it from a Mongo shell. The monitoring agent, used to send the statistics of the Mongo instance to the monitoring service, uses Python and PyMongo. Refer to the *Installing PyMongo* recipe in *Chapter 3, Programming Language Drivers*, to know more about how to install Python and PyMongo, the Python client of MongoDB.

How to do it...

1. If you don't already have an MMS account, then log in at https://mms.mongodb.com/ and sign up for an account. On signing up and logging in, we should see the following page:

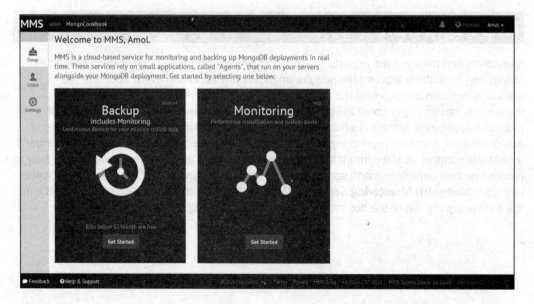

Chapter 6

2. Click on the **Get Started** button under **Monitoring**.
3. Once we reach the **Download Agent** option in the menu, click on the appropriate OS platform to download the agent. Follow the instructions given, after selecting the appropriate OS platform. Note down the API key too. For example, if the Windows platform is selected, we would see the following page:

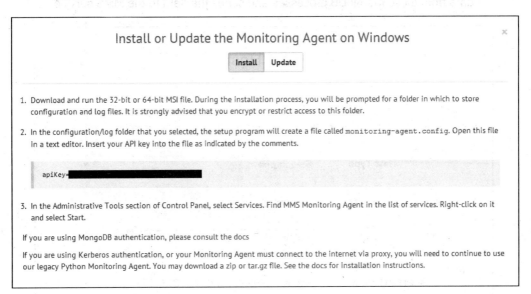

4. Once the installation is complete, open the `monitoring-agent.config` file, which will be present in the configuration folder selected while installing the agent.
5. Look out for the `mmsApiKey` key in the file and set its value to the API key that was noted down earlier in step 3.
6. To start a service manually, we have to go to `services.msc` on MS Windows, which can be done by typing `services.msc` in the *Run* dialog (Windows + *R*). The service will be named MMS Monitoring Agent. On the web page, click on the **Verify Agent** button. If all goes well, the started agent will be verified and the success message will be shown.

Monitoring and Backups

7. The next step is to configure the host. This host is the one that is seen from the agent's perspective, running on the organization/individual's infrastructure. The following screen shows the screen used for the addition of a host. The hostname is the internal hostname (the hostname on the client's network); the MMS on the cloud doesn't need to reach out to the MongoDB processes. It is the agent that collects the data from these MongoDB processes and sends the data to the MMS service.

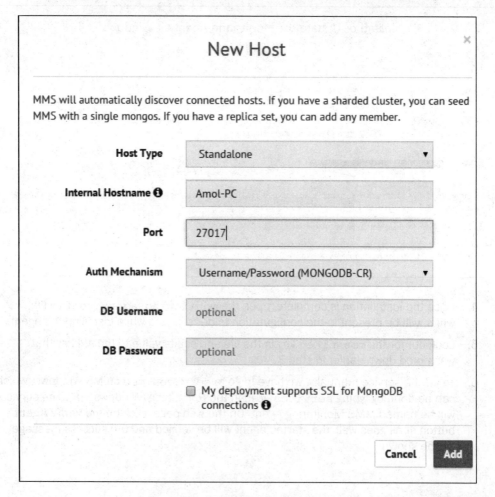

8. Once the host details are added, click on the **Verify Host** button. Once the verification is done, click on the **Start Monitoring** button.

We have successfully set up MMS and added one host to it, which would be monitored.

How it works...

In this recipe, we have set up the MMS agent and monitoring for a standalone MongoDB instance. The installation and setup process is pretty simple. We also added a standalone instance and all was ok.

Suppose we have a replica set up and running (refer to the *Starting multiple instances as part of a replica set* recipe in *Chapter 1, Installing and Starting the MongoDB Server*, for more details on how to start a replica set), and the three members are listening to ports `27000`, `27001`, and `27002`, respectively. Refer to step 7 in the *How to do it...* section, where we set up one standalone host. If we select **Replica Set** in the dropdown for **Host Type**, and for the internal hostname we give a valid hostname of any member of the replica set (in my case, `Amol-PC` and port `27001` were given, which is a secondary instance), all other instances will automatically be discovered and they will be visible under the hosts, as shown in the following screenshot:

We didn't see what is to be done when security is enabled on the cluster, which is pretty common in production environments. If authentication is enabled, we need proper credentials for the MMS agent to gather the statistics. The DB username and password that we give while adding a new host (step 7 of the *How to do it...* section) should have a minimum of `clusterAdmin` and `readAnyDatabase` roles.

There's more...

What we saw in this recipe was setting up an MMS agent and creating an account from the MMS console. However, we can add groups and users for the MMS console as administrators, granting various users privileges for performing various operations on different groups. In the next recipe, we will throw some light on user and group management in the MMS console.

Monitoring and Backups

Managing users and groups on the MMS console

In the previous recipe, we saw how to set up an MMS account and how to set up an MMS agent. In this recipe, we will throw some light on how to set up the groups and user access to the MMS console.

Getting ready

Refer to the previous recipe for setting up the agent and the MMS account. This is the only prerequisite for this recipe.

How to do it...

1. Start by navigating to **Administration** | **Users** on the left-hand side of the screen, as shown in the following screenshot:

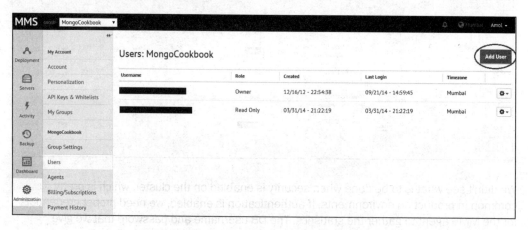

Chapter 6

2. Here you can view the existing users and also add new users. On clicking on the **Add User** button (circled in the top-right corner of the previous screenshot), you should see the following pop-up window allowing you to add a new user:

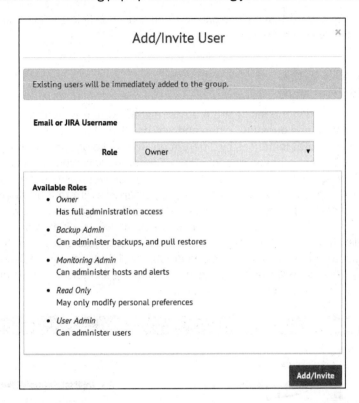

The preceding screen will be used to add users. Take note of the various available roles.

3. Similarly, by navigating to **Administration | My Groups**, you can view and also add new groups, by clicking on the **Add Group** button. In the textbox, provide a name for the group. Remember that the name of the group you enter should be available globally. The given name of the group should be unique across all user bases of MMS and not just your account.

When a new group is created, it will be visible in the upper-left corner in a dropdown for all the groups, as shown in the following screenshot:

Monitoring and Backups

You can switch between the groups using this dropdown, which should show all the details and stats relevant to the selected group.

 Remember that a group once created cannot be deleted. So be careful while creating one.

How it works...

The tasks we completed in the recipe are pretty straightforward and don't need a lot of explanation, except for one question. When and why do we add a group? It is when we want to segregate our MongoDB instances by different environments or applications. There will be a different MMS agent running for each group. Creating a new group is necessary when we want to have separate monitoring groups for different environments of an application (development, QA, production, and so on), and each group has different privileges for the users. That is, the same agent cannot be used for two different groups. If we remember from the previous recipe, while configuring the MMS agent, we give it an API key unique to the group. To view the API key for the group, select the appropriate group from the dropdown on the top (if your user has access only to one group, the dropdown won't be seen), go to **Administration** | **Group Settings**, as shown in the following screenshot. The group ID and the API key will both be shown at the top of the page.

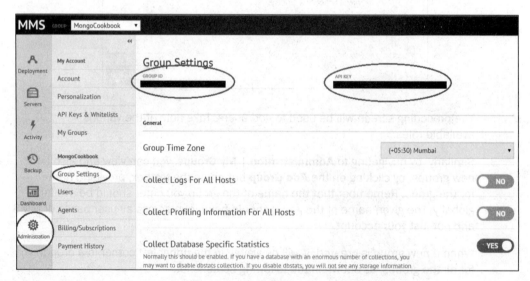

Note that not all user roles will see this option. For example, users with read-only privileges can only personalize their profile, and most of the other options will not be visible.

Chapter 6

Monitoring MongoDB instances on MMS

The previous recipes, *Signing up for MMS and setting up the MMS monitoring agent* and *Managing users and groups in the MMS console*, showed us how to set up an MMS account and agent, add hosts, and manage user access to the MMS console. The core objective of MMS is monitoring the host instances, which is still not discussed. In this recipe, we will be performing some operations on the host that we added to MMS in the first recipe, and we will monitor it from the MMS console.

Getting ready

Follow the recipe *Signing up for MMS and setting up the MMS monitoring agent* and that is pretty much what is needed for this recipe. You may choose to have a standalone instance or a replica set, either ways is fine. Also, open a Mongo shell and connect to the primary instance from it (it is a replica set).

How to do it…

1. Start by logging into the MMS console and clicking on **Deployment** in the upper-left corner, and then again on the **Deployment** link in the submenu, as shown in the following screenshot:

2. Clicking on one of the hostnames shown, we will see a large variety of graphs showing various statistics. In this recipe, we will analyze a majority of these.

3. Open the bundle downloaded for the book. In *Chapter 4, Administration*, we used a JavaScript file named `KeepServerBusy.js` to keep the server busy with some operations. We will be using the same script this time around.

Monitoring and Backups

4. In the operating system shell, execute the following command with the `.js` file in the current directory. The shell connects to the port, in my case port 27000, for the primary.

   ```
   $ mongo KeepServerBusy.js --port 27000 --quiet
   ```

5. Once started, keep it running and give it 5-10 minutes before you start monitoring the graphs on the MMS console.

How it works...

The *Understanding the mongostat and mongotop utilities utilities* recipe in *Chapter 4, Administration*, demonstrated how these utilities can be used to get the current operations and resource utilization. That is a fairly basic and helpful way to monitor a particular instance. MMS, however, gives us one place to monitor the MongoDB instance with pretty easy-to-understand graphs. MMS also gives us historical stats, which `mongostat` and `mongotop` cannot give.

Before we go ahead with the analysis of the metrics, I would like to mention that in case of MMS monitoring, the data is not queried nor sent out over the public network. It is just the statistics that are sent over a secure channel by the agent. The source code for the agent is open source and is available for examination if needed. The `mongod` servers need not be accessible from the public network, as the cloud-based MMS service never communicates to the server instances directly. It is the MMS agent that communicates to the MMS service. Typically, one agent is enough to monitor several servers, unless you plan to segregate them into different groups. Also, it is recommended to run the agent on a dedicated machine / virtual machine and not share it with any of the `mongod` or `mongos` instances, unless it is a less crucial test instance group you are monitoring.

Let us see some of these statistics on the console; we start with the memory related ones. The following graph shows the resident, mapped, and virtual memory:

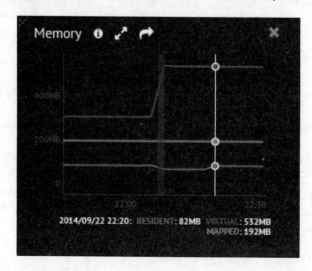

As seen in the previous graph, the resident memory for the data set is 82 MB, which is very low, and it is the actual physical memory used up by the mongod process. This current value is significantly below the free memory available, and generally, this will increase over a period of time until it reaches a point where it has used up a large chunk of the total available physical memory. This is automatically taken care of by the mongod server process, and we can't force it to use up more memory, even though it is available on the machine it is running on.

The mapped memory, on the other hand, is about the total size of the database, and is mapped by MongoDB. This size can be (and usually is) much higher than the physical memory available, which enables the mongod process to address the entire dataset as it is present in memory even if it isn't present. MongoDB offloads this responsibility of mapping and loading of data to and from the disk to the underlying operating system. Whenever a memory location is accessed and it is not available in the RAM (that is, the resident memory), the operating system fetches the page into memory, evicting some page to make space for the new page if necessary. What exactly is a memory-mapped file? Let us try to see with a super-scaled-down version. Suppose we have a file of 1 KB (1024 bytes) and the RAM is only 512 bytes, then obviously we cannot have the whole file in the memory. However, you can ask the operating system to map this file to the available RAM in pages. Suppose the page is of 128 bytes, then the total file is eight pages (128 * 8 = 1024). However, the OS can load four pages only, and assume that it loaded the first four pages (up to 512 bytes) in memory. When we access the byte number 200, it is ok and found in memory, as it is present on page 2. But what if we access byte 800, which is logically on page 7, which is not loaded in memory? What the OS does is, it takes one page out from the memory and loads page 7, which contains byte number 800. MongoDB as an application gets a feel that everything was loaded in memory and was accessed by the byte index, but actually it wasn't, and OS transparently did the work for us. As the page accessed was not present in memory and we had to go to the disk to load it in memory, it is called a page fault.

Getting back to the stats shown in the graph, the virtual memory contains all the memory usage, including the mapped memory, plus any additional memory used, such as the memory associated with the thread stack associated with each connection, and so on. If journaling is enabled, this size will definitely be more than twice that of the mapped memory, as journaling too will have a separate memory mapping for the data. Thus we have two addresses mapping the same memory location. This doesn't mean that the page will be loaded twice. It just means that two different memory locations can be used to address the same physical memory. Very high virtual memory might need some investigations. There is no predetermined value for what too high or a low value is; generally these values are monitored for your system under normal circumstances when you are happy with the performance of your system. These benchmark values should then be compared with the figures seen when the system performance goes down, and then appropriate actions can be taken.

As we saw earlier, page faults are caused when an accessed memory location is not present in the resident memory, causing OS to load the page from the memory. This IO activity will definitely cause the performance to reduce, and too many page faults can bring down the database performance dramatically. The following graph shows quite a few page faults occurring per minute. However, if the disk used is SSDs instead of the spinning disk, the hit in terms of seek time from drive might not be significantly high.

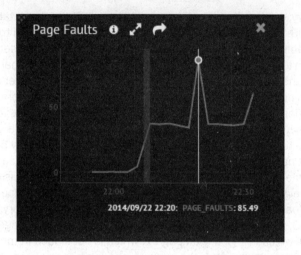

A large number of page faults usually occur when enough physical memory isn't available to accommodate the data set, and the operating system needs to get the data from the disk into the memory. Note that this stat shown earlier is taken on an MS Windows platform and this graph might seem high for a very trivial operation. The value shown here is the sum of hard and soft page faults and doesn't really give a true figure of how good (or bad) the system is doing. These figures would be different on a Unix-based operating system. There is a JIRA open at the time of writing this book, which reports this problem (`https://jira.mongodb.org/browse/SERVER-5799`).

One thing you might need to remember is that, in production systems, MongoDB doesn't work well with NUMA architecture and you might see a lot of page faults occurring even if the available memory seems to be high enough. Refer to `http://docs.mongodb.org/manual/administration/production-notes/` for more details.

There is an additional graph, as seen next, which gives some details about nonmapped memory. As we saw earlier in this section, there are three types of memory, namely, mapped, resident, and virtual. Mapped memory is always less than virtual memory. Virtual memory will be more than twice that of mapped memory if journaling is enabled. If we look at the graph given earlier in this section, we see that the mapped memory is 192 MB, whereas the virtual memory is 532 MB. As journaling is enabled, the memory is more than twice that of the mapped memory. When journaling is enabled, the same page of data is mapped twice in memory. Note that the page is physically loaded only once; it is just that the same location can be addressed using two different addresses.

Let us find the difference between the virtual memory, which is 532 MB and twice the mapped memory, which is *2 * 192 = 384* MB. The difference between these figures is 148 MB (*532 - 384*).

What we see next is the portion of virtual memory that is not mapped. This value is the same as what we just calculated.

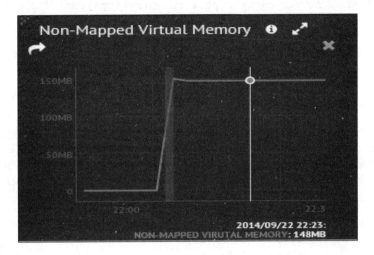

As mentioned earlier, a high or low value for nonmapped memory is not defined; however, when the value reaches GBs, we might have to investigate, if the possible number of open connections is high, and check if there is a leak with client applications not closing them after using it. There is a graph that gives us the number of connections open and it looks as follows:

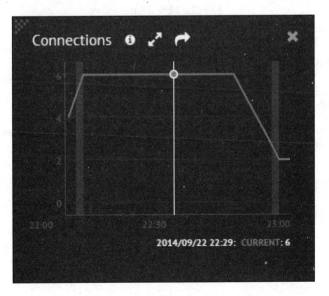

Monitoring and Backups

Once we know the number of connections and find it too high as compared to the normal expected count, we will need to find the clients who have opened the connections to that instance. We can execute the following JavaScript code from the shell to get those details. Unfortunately, at the time of writing this book, MMS didn't have this feature to list out the client connection details.

```
testMon:PRIMARY> var currentOps = db.currentOp(true).inprog;
currentOps.forEach(function(c) {
    if(c.hasOwnProperty('client')) {
          print('Client: ' + c.client + ", connection id is: " + c.desc);
     }
    //Get other details as needed
});
```

The `db.currentOp` method returns all the idle and system operations in the result. We then iterate through all the results and print out the client host and the connection details. A typical document in the result of the `currentOp` method looks like the following code snippet. You may choose to tweak the preceding piece of code to include more details according to your needs.

```
{
              "opid" : 62052485,
              "active" : false,
              "op" : "query",
              "ns" : "",
              "query" : {
                      "replSetGetStatus" : 1,
                      "forShell" : 1
              },
              "client" : "127.0.0.1:64460",
              "desc" : "conn3651",
              "connectionId" : 3651,
              "waitingForLock" : false,
              "numYields" : 0,
              "lockStats" : {
                      "timeLockedMicros" : {

                      },
                      "timeAcquiringMicros" : {

                      }
              }
}
```

Chapter 6

The *Understanding the mongostat and mongotop utilities* recipe in *Chapter 4, Administration*, was used to get some details on the percentage of time for which a database was locked, and the number of `update`, `insert`, `delete`, and `getmore` operations executed per second. You may refer to this recipe and try it out. We used the same JavaScript that we have used currently to keep the server busy.

In the MMS console, we have similar graphs giving these details as follows:

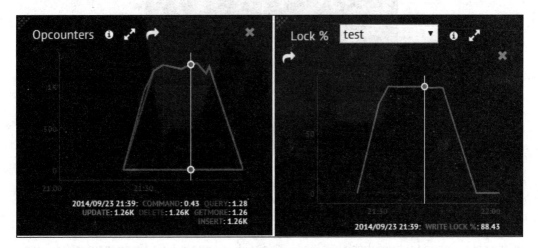

The first one, **Opcounters**, shows the number of operations executed as of a particular point in time. This should be similar to what we saw using the `mongostat` utility. Similarly, the one on the right shows us the percentage of time for which a DB was locked. The previous dropdown lists out the database names; we can select an appropriate database for which we want to see the stats. Again, this statistic can be seen using the `mongostat` utility. The only difference is, with the command-line utility, we see the stats as of the current time whereas here, we see the historical stats as well.

In MongoDB, indexes are stored in B-trees, and the following graph shows the number of times the B-tree index was accessed, hit, and missed. At the minimum, the RAM should be enough to accommodate the indexes for optimum performance; so in metrics, the misses should be zero or very low. A high number of misses results in a page fault for the index and possibly, additional page faults for the corresponding data, if the query is not covered; all its data cannot be sourced from the index, which is a double blow for its performance. One good practice, whenever querying, is to use projections and fetch only the necessary fields from the document. This is helpful whenever we have our selected fields present in an index, in which case, the query is covered and all the necessary data is sourced only from the index.

To find out more about covered indexes, refer to the *Creating an index and viewing plans of queries* recipe in *Chapter 2, Command-line Operations and Indexes*.

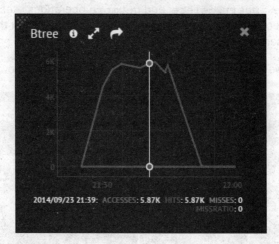

For busy applications, when MongoDB acquires a lock on the database, other read and write operations get queued up. If the volumes are very high with multiple write and read operations contending for lock, the operations queue up. Until version 2.4 of MongoDB, the locks are at database level; thus, even if the writes are happening on another collection, read operations on any collection in that database will block. This queuing operation affects the performance of the system and is a good indicator that the data might need to be sharded across to scale the system.

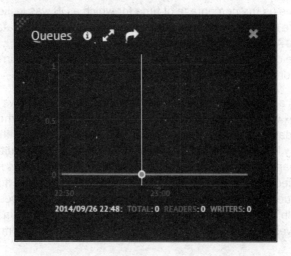

 Remember, no value is defined as high or low; it is an acceptable value based on an application to application basis.

MongoDB flushes the data immediately from the journal and periodically from the data file to the disk. The following metrics give us the flush time per minute at a given point in time. If the flush takes up a significant percentage of the time per minute, we can safely say that the write operations are forming a bottleneck for the performance.

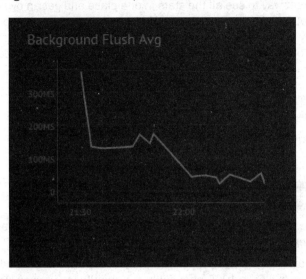

There's more...

We have seen monitoring of the MongoDB instances/cluster in this recipe. However, setting up alerts to be notified when certain threshold values are crossed, is what we still haven't seen. In the next recipe, we will see how to achieve this with a sample alert, which is sent out over an e-mail when the page faults cross a predetermined value.

See also

- Monitoring hardware, such as CPU usage, is pretty useful, and the MMS console does support that. It, however, needs munin-node to be installed to enable CPU monitoring. Refer to http://mms.mongodb.com/help/monitoring/configuring/ to set up munin-node and hardware monitoring.
- To update the monitoring agent, refer to http://mms.mongodb.com/help/monitoring/tutorial/update-mms/.

Monitoring and Backups

Setting up monitoring alerts on MMS

In the previous recipe, we saw how we can monitor various metrics from the MMS console. This is a great way to see all the stats in one place and get an overview of the health of the MongoDB instances and cluster. However, it is not possible to monitor the system continuously for the support personnel, and there has to be some mechanism to automatically send out alerts in the case of some threshold being exceeded. In this recipe, we will set up an alert whenever the page faults exceed 1000.

Getting ready

Refer to the *Monitoring MongoDB instances on MMS* recipe. This is the only prerequisite for this recipe.

How to do it...

1. Click on the **Activity** option from the left-hand side menu options and then click on **Alert Settings**. On the **Alert Settings** page, click on **Add Alert**.
2. Add a new alert for the host, which is a primary instance, if the page faults exceed a given number, which is 1000 page faults per minute in our case. The notification was chosen to be e-mail in this case, and the interval after which the alert will be sent is set at 10 minutes.

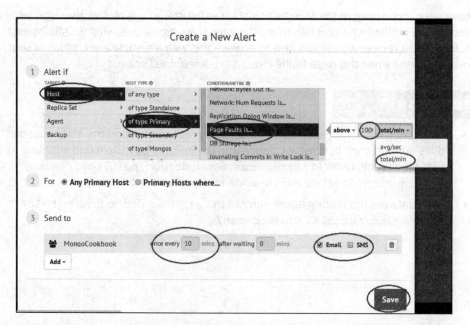

3. Click on **Save** to save the alert.

Chapter 6

How it works...

The steps were pretty simple. What we did was successfully set up MMS alerts when the page faults exceeded 1000 per minute. As we saw in the previous recipe, no fixed value is classified as high or low. It is something that is acceptable for your system, which should come with benchmarking the system during the testing phases in your environment. Similar to page faults, there is a vast array of alerts that can be set up. Once an alert is raised, it will be sent every 10 minutes, as we have set until the condition for sending the alerts is not met, which, in this case, is if the number of page faults fall below 1000 or somebody manually acknowledges the alert which means no alert will be sent further for that incident.

As we see in the following screenshot, the alert is open and we can acknowledge the alert:

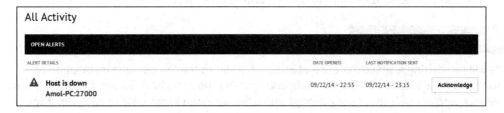

On clicking **Acknowledge**, the following pop up will let us choose the duration for which we will acknowledge:

This means that for this particular incident, no more alerts will be sent out until the selected time period elapses.

The open alerts can be viewed by clicking on the **Activities** menu option on the left-hand side of the page.

Monitoring and Backups

See also

- Visit `http://www.mongodb.com/blog/post/five-mms-monitoring-alerts-keep-your-mongodb-deployment-track` for some of the important alerts that you should set up for your deployment

Backing up and restoring data in Mongo using out-of-the box tools

In this recipe, we will look at some basic backup and restore operations using utilities such as `mongodump` and `mongorestore` to backup and restore files.

Getting ready

We will be starting a single instance of `mongod`. Refer to the *Single node installation of MongoDB* recipe in *Chapter 1, Installing and Starting the MongoDB Server*, to start a Mongo instance and connect to it from a Mongo shell. We will need some data to back up; if you already have some data in your test database that would be fine, else create some from the `countries.geo.json` file available in the code bundle, using the following command:

```
$ mongoimport  -c countries -d test --drop countries.geo.json
```

How to do it...

1. With the data in the test database, execute the following command, assuming we want to export the data to a local directory called `dump` in the current directory:

    ```
    $ mongodump -o dump -oplog -h localhost -port 27017
    ```

    Verify that there is data in the `dump` directory. All files should be `.bson` files, one per collection, in the respective database folder created.

2. Now let us import the data back into the MongoDB server using the following command. This is again with an assumption that we have the directory dump in the current directory with the required `.bson` files present in it.

    ```
    mongorestore --drop -h localhost -port 27017 dump -oplogReplay
    ```

How it works...

We executed just a couple of steps to export and restore the data. Let us now see exactly what it does and what the command-line options for this utility are. The `mongodump` utility is used to export the database into `.bson` files, which can later be used to restore the data in the database. The export utility exports one folder per database, except the local database, and then each of them will have one `.bson` file per collection. In our case we used the `-oplog` option to export a part of the oplog as well, and the data will be exported to the `oplog.bson` file. Similarly, we import the data back into the database using the `mongorestore` utility. We explicitly ask the existing data to be dropped by providing the `--drop` option before the import and replay of the contents in the oplog, if any.

The `mongodump` utility simply queries the collection and exports the contents to the files. The bigger the collection, the more will be the time taken to restore the contents. It is thus advisable to prevent the write operations when the dump is being taken. In case of sharded environments, the balancer should be turned off. If the dump is taken while the system is running, export it with the `-oplog` option to export the contents of the oplog as well. This oplog can then be used to restore the point-in-time data. The following are some of the important options available for the `mongodump` and `mongorestore` utilities, first for `mongodump`.

Option	Description
`--help`	This shows all the possible supported options and a brief description of those options.
`-h` or `--host`	This is the host that must be connected to. By default, it is localhost on port `27017`. If a standalone instance is to be connected to, we can give the hostname as `<hostname>:<port number>`. For a replica set, the format will be `<replica set name>/<hostname>:<port>,....<hostname>:<port>`, where the comma-separated list of hostnames and ports is called the seed list, which can contain all or a subset of hostnames in a replica set.
`--port`	This is the port number of the target MongoDB instance. It is not really relevant if the port number is provided in the previous `-h` or `--host` option.
`-u` or `--username`	This provides the username of the user, using which the data would be exported. As the data is read from all databases, the user is at least expected to have read privileges in all databases.
`-p` or `--password`	This is the password used in conjunction with the username.

Monitoring and Backups

Option	Description
`--authenticationDatabase`	This is the database in which the user credentials are kept; if not specified, the database specified in the `--db` option is used.
`-d` or `--db`	This is the database to backup. If not specified, then all the databases are exported.
`-c` or `--collection`	This is the collection in the database to be exported.
`-o` or `--out`	This is the directory to which the files will be exported. By default, the utility will create a dump folder in the current directory and export the contents to that directory.
`--dbpath`	The value is the directory where the database files will be found. Use this option only when we intend not to connect to a running MongoDB instance but write to the database files directly. The server should not be up and running while reading directly from the database files, as the export locks the data files, which can't happen if a server is up and running. A lock file will be created in the directory while the lock is acquired.
`--oplog`	With the option enabled, the data from the oplog from the time the export process started is also exported. Without this option enabled, the data in the export will not represent a single point in time if writes are happening in parallel, as the export process can take few hours and it simply is a query operation on all the collections. Exporting the oplog gives an option to restore a point-in-time data. There is no need to specify this option if you are preventing write operations while the export is in progress.

Similarly, for the `mongorestore` utility, the options are as follows. The meaning of the options `--help`, `-h` or `--host`, `--port`, `-u` or `--username`, `-p` or `--password`, `--authenticationDatabase`, `-d` or `--db`, `-c` or `-collection` is same as in case of mongodump:

Option	Description
`--dbpath`	The value is the directory where the database files will be found. Use this option only when we intend not to connect to a running MongoDB instance but write to the database files directly. The server should not be up and running while writing directly to the database files, as the restore operation locks the data files, which can't happen if a server is up and running. A lock file will be created in the directory while the lock is acquired.
`--drop`	Drop the existing data in the collection before restoring the data from the exported dumps.

Option	Description
`--oplogReplay`	If the data was exported while writes to the database were allowed, and if the `--oplog` option was enabled during export, the oplog exported will be replayed on the data to bring the entire data in the database to the same point in time.
`--oplogLimit`	The value of this parameter is a number representing the time in seconds. This option is used in conjunction with the `oplogReplay` command-line option, which is used to tell the restore utility to replay the oplog and stop just at the limit specified by this option.

One might even think "why not copy the files and take a backup?". That works well, but there are a few problems associated with it. The first being, you cannot get a point-in-time backup unless the write operations are disabled and secondly, the space used for backups is very high, as the copy would also copy the zero-padded files of the database, as against the `mongodump` utility that exports just the data.

Having said that, filesystem snapshotting is a commonly used practice for backups. One thing to remember is that, while taking the snapshot, the journal files and the data files need to come in the same snapshot for consistency.

Configuring the MMS backup service

MMS backup is a relatively new offering by MongoDB for real-time incremental backup of your MongoDB instances, replica sets, and shards, and it offers you point-in-time recovery for your instances. The service is available as on-prem (in your data center) or cloud. We will, however, be demonstrating the on-cloud service, which is the only option for the community and basic subscriptions. For more details on the available options, you can refer to the different product offerings by MongoDB at https://www.mongodb.com/products/subscriptions.

Getting ready

The Mongo MMS backup service will work only on Mongo 2.0 and above. We will start a single server that we would backup. MMS backup relies on the oplog for continuous backup, and as oplog is available only in replica sets, the server needs to be started as a replica set. Refer to the *Installing PyMongo* recipe in *Chapter 3, Programming Language Drivers*, to know more about how to install Python and the Python client of Mongo, PyMongo.

Monitoring and Backups

How to do it...

1. If you don't have an MMS account already, then log in to `https://mms.mongodb.com/` and sign up for an account. For screenshots, refer to the *Signing up for MMS and setting up the MMS monitoring agent* recipe.

2. Start a single instance of Mongo by replacing the value of the appropriate filesystem path on your machine as follows:

   ```
   $ mongod --replSet testBackup --smallfiles --oplogSize 50 --dbpath /data/mongo/db
   ```

   Note that `smallfiles` and `oplogSize` are options set only for the purpose of testing and they are not to be used in production.

3. Start a shell, connect to this started instance, and initiate the replica set as follows:

   ```
   > rs.initiate()
   ```

   The replica set will be up and running in some time.

4. Go back to the browser and point to `mms.mongodb.com`. Add a new host by clicking on the **+ Add Host** button. Select the type as replica set and `hostname` as your hostname and the default port (`27017` in our case). Refer to the *Signing up for MMS and setting up the MMS monitoring agent* recipe for the screenshots of the add host process.

5. Once the host is successfully added, register for MMS backup by clicking on the **Backup** option on the left-hand side and then on **Begin Setup**.

6. An SMS or Google Authenticator can be used for registration. If a smartphone is available with Android, iOS, or Blackberry OS, Google Authenticator is a good option. For some countries such as India, Google Authenticator is the only option available.

7. Assuming Google Authenticator is not configured already and we are planning to use it, we would need the app to be installed on your smartphone. Go to the respective app store of your mobile OS platform and install the Google Authenticator software.

8. With the software installed on the phone, come back to the browser. We should see the following screen on selecting Google Authenticator:

Chapter 6

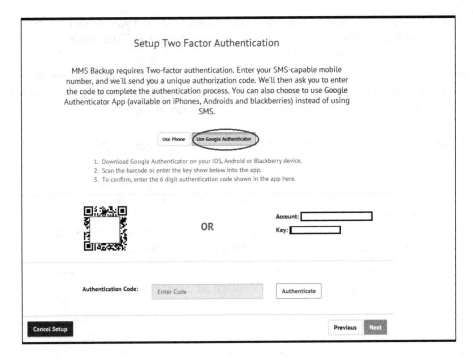

9. Begin the set up for a new account by scanning the QR code from the Google Authenticator application. If barcode scanning is a problem, you may choose to manually enter the key given on the right-hand side of the screen.

10. Once the scanning is completed or the key is entered successfully, your smartphone should show a six-digit number that changes every 30 seconds. Enter that number in the **Authentication Code** box given on the screen.

> It is important not to delete this account in Google Authenticator on your phone, as this would be used in future whenever we wish to change any settings related to backup, such as stopping backup, changing the exclusion list, and literally any operation in MMS backup. The QR code and key would not be visible again once the setup is done. You would have to contact MongoDB support to get the configuration reset.

11. Once the authentication is done, the next screen you should see is for the billing address and billing details, such as the card you register. All charges below USD 5 are waived off, so you should be ok to try out a small test instance before being charged.

12. Once the credit card details are saved, we move ahead with the setup. We will have to install a backup agent; this is a separate agent from the monitoring agent. Choose the appropriate platform and follow the instructions for its installation. Take note of the location where the configuration files of the agent will be placed.

Monitoring and Backups

13. A new pop up will contain the instructions/link to the archive/installer for the platform and the steps to install. It should also contain the `apiKey`. Take note of that API key, which we will need in the next step.

14. Once the installation is complete, open the `local.config` file placed in the `config` directory of the agent installation (the location that was shown/modified during the installation of the agent) and paste/type in the `apiKey` noted down in the previous step.

15. Once the agent is configured and started, click on the **Verify Agent** button.

16. Once the agent is successfully verified, we should start by adding a host to back up. The dropdown should show us all the replica sets and shards we have added. Select the appropriate one and fill the sync source as the primary instance, as that is the only one we have in our standalone instance. Sync source is only used for the initial sync process. Whenever we have a proper replica set with multiple instances, it is preferable to use a secondary as a sync-process instance.

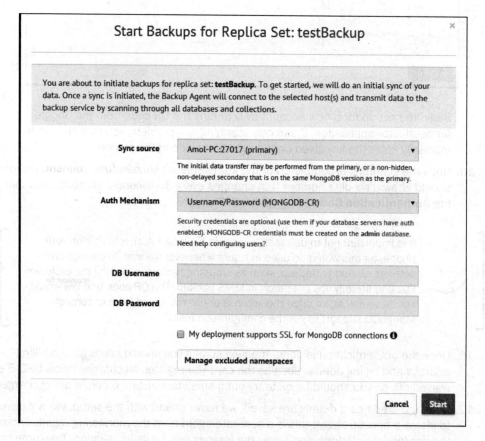

As the instance is not started with security, leave the **DB Username** and **Password** fields blank.

Chapter 6

17. Click on the **Manage excluded namespaces** button if you wish to skip a particular database or collection being backed up. If nothing is provided, by default, everything will be backed up. The format for the collection name would be `<databasename>.<collection name>`. Alternatively, it could be just the database name, in which case all collections in that database would not be eligible for backup.
18. Once the details are all ok, click on the **Start** button. This should complete the setup of the backup process for a replica set on MMS.

> The installation steps I performed were on Windows OS, and the service needs to be started manually in that case. Press the Windows button + R and type `services.msc`. The name of the service is MMS Backup Agent.

How it works...

The steps are pretty simple and this is all we need to do to set up a server for Mongo MMS backup. One important thing mentioned earlier is that MMS backup uses multifactor authentication for any operation once the backup is set up; the account setup in Google Authenticator for MongoDB should not be deleted. There is no way to recover the original key used to set up the authenticator. You will have to clear the Google Authenticator settings and set up a new key. To do that, click on the **Help & Support** link at the bottom left of the screen and click on **How do I reset my two-factor authentication?**.

On clicking the link, a new window will open up, as shown in the following screenshot, which will ask for the username. An e-mail will be sent out to the registered e-mail ID, which allows you to reset the two-factor authentication.

FAQ

General

How do I reset my password?
How do I reset my two-factor authentication?
How do I remove my company/group?
How do I modify alerts?
What services does MongoDB Support provide?

View All

Reset Two Factor Authentication

In order to reset your two-factor authentication, you must be able to receive email at the address associated with your account, know your password, and know the **Agent API Key** for any Group of which you are a member. The Agent API Key may be found in the configuration file for your Monitoring or Backup Agent.

Username

[Reset Two Factor Authentication]

Monitoring and Backups

As mentioned, oplog is used to synchronize the current MongoDB instance with the MMS service. However, for the initial sync, an instance's data files are used. Which instance to use is provided by us when we set up the backup of the replica set. As this is a resource-heavy operation, we must preferably use a secondary instance for this on busy systems so as not to add more querying on the primary instance by the MMS backup agent. Once the instance is done with initial synchronization, the oplog of the primary will be used to get data on a continuous basis. The agent does write periodically to a collection called `mms.backup` in the admin database.

The backup agent for MMS backup is different from the MMS monitoring agent. Though there is no restriction on having them both run on the same machine, you might need to evaluate that before having such a setup in production. A safe bet would be to have them running on separate machines. Never run either of these agents with a `mongod` or `mongos` instance on the same box in production. There are a couple of important reasons why it is not recommended to run the agents on the same box as the `mongod` instances. They are as follows:

- The resource utilization of the agent is dependent on the cluster size it monitors. We don't want the agent to use a lot of resources affecting the performance of the production instance.
- The agent could be monitoring a lot of server instances at one time. As there is only one instance of this agent, we do not want it to go down during the database server maintenance and restart.

The community edition of MongoDB built with SSL or the Enterprise versions, with the SSL option used for communication between the client and the MongoDB server, must perform some additional steps. The first step is to check the **My deployment supports SSL for MongoDB connections** flag when we set up the replica set for backup (see step 16). Note the checkbox at the bottom of the screenshot; it should be checked. Secondly, open the `local.config` file for the MMS configuration and look out for the following two properties:

- `sslTrustedServerCertificates=`
- `sslRequireValidServerCertificates=true`

The first is a fully-qualified path of the certifying authority's certificate in the PEM format. This certificate will be used to verify the certificate presented by the `mongod` instance running over SSL. The second property can be set to `false` if the certificate verification is to be disabled; this is however not a recommended option. As far as the traffic between the backup agent and MMS backup is concerned, data sent from the agent to the MMS service over SSL is secure, irrespective of whether SSL is enabled on your MongoDB instances or not. The data at rest in the data center for the backed up data is not encrypted.

If security is enabled on the `mongod` instance, a username and password needs to be provided, which will be used by the MMS backup agent. The username and password are provided while setting up backup for the replica set, as seen in step 16.

As the agent needs to read the oplog, possibly all databases for the initial sync and write data to the admin database; the roles expected from the user are `readAnyDatabase`, `clusterAdmin`, `readWrite` on admin and local database, and `userAdminAnyDatabase` database role in the case of version 2.4 and above. In versions prior to 2.4, we would expect the user to have read access on all the databases and read/write access to admin and local databases.

While setting up a replica set for backup you may get an error such as, **Insufficient oplog size: The oplog window must be at least 1 hours over the last 24 hours for all active replica set members. Please increase the oplog**. While you may think this is always something to do with oplog size, it is also seen when the replica set has an instance that is in a recovery state. This might feel misleading, so do look out for recovering nodes, if any, in the replica set, while setting up a backup for a replica set. As per the MMS support, it seems too restrictive to not let set up a replica set for backup with some recovering nodes and it might be fixed in future.

Managing backups in the MMS backup service

In the previous recipe, we learned how to set up the MMS backup service and a simple one-member replica set was set up for backup. Though a single member replica set makes no sense at all, it was needed, as a standalone instance cannot be set up for backup in MMS. In this recipe, we dive deeper and look at the operations we can perform on the server that is set up for backup, such as starting, stopping, or terminating a backup; managing exclusion lists; managing backup snapshots; and retaining and restoring to point-in-time data.

Getting ready

The previous recipe is all that is needed to be followed for this recipe. The necessary setup described in it is expected to be done, as we are going to use the same server we had set up for backup in that recipe.

How to do it...

1. With the server up and running, let's import some data in it. It can be anything, but we chose to use the `countries.geo.json` file that was used in the previous chapter. It should be available in the bundle downloaded from the Packt Publishing website.

    Start by importing the data into a collection called `countries` in the test database. Use the following command to do it. The following import command was executed with the current directory having the `countries.geo.json` file:

    ```
    $ mongoimport  -c countries -d test --drop countries.geo.json
    ```

Monitoring and Backups

2. We have already seen how to exclude namespaces when the replica set backup was being set up. We will now see how to exclude namespaces once the backup for a replica set is done. Click on the **Backup** menu option on the left and then on **Replica Set Status**, which opens by default when **Backup** is clicked. Click on the gear button on the right-hand side of the row where the replica set is shown. It should look as follows:

3. As shown in the previous screenshot, click on the **Edit Excluded Namespaces** option and type in the name of the collection that we want to exclude. Suppose we want to exclude the `applicationLogs` collection in the test database, type in `test.applicationLogs`.

4. On saving it, you will be asked to enter the token code that is currently displayed on your Google Authenticator.

5. On successful validation of the code, the namespace `test.applicationLogs` will be added to the list of namespaces excluded from being backed up.

6. We shall now see how to manage the snapshot scheduling. Snapshot is the state of the database as of a particular point in time. To manage the snapshot frequency and retention policy, click on the gear button shown in step 2 and click on **Edit Snapshot Schedule**.

7. As seen in the following screenshot, we can set the times when the snapshots are taken and their retention period. More on this will be covered in the next section. Any changes to it would need multifactor authentication to save the changes.

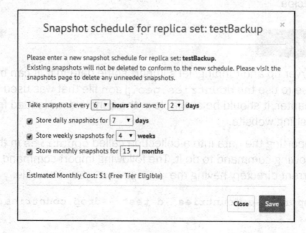

Chapter 6

8. We will now look at how we go about restoring the data using MMS backup. At any point in time whenever we want to restore the data, click on **Backup** and the **Replica Set Status/Shard Cluster Status** and then click on the set/cluster name.

9. On clicking it, we will see the snapshots that are saved against this set. It should look something like what is seen in the following screenshot:

We have encircled some of the portions on the screen, which we will see one by one.

10. To restore as of a time when the snapshot was taken, click on the **Restore this snapshot** link in the **ACTIONS** column of the grid.

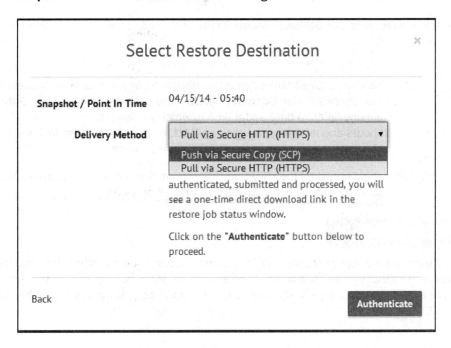

261

Monitoring and Backups

11. The previous screenshot shows us how we can export the data, either over HTTPS or SCP. We select **Pull via Secure HTTP (HTTPS)** for now, and click on **Authenticate**. We will see about SCP in the next section.

12. Enter the token that is received either over SMS or seen on Google Authenticator, and click on **Finalize Request** on entering the auth code.

13. On successful authentication, click on **Restore Jobs** as shown in the following screenshot. This is a one-time download that will let you download the `tar.gz` archive. Click on the download link to download the `tar.gz` archive.

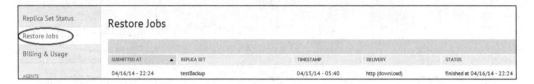

14. Once the archive is downloaded, extract it to get the database files within it.

15. Stop the `mongod` instance, replace the database files with the ones that are extracted, and restart the server to get the data as of the time when the snapshot was taken. Note that the database file will not contain data for the collection that was excluded from backup if at all.

We will now see how to get the point-in-time data using MMS backup:

1. Click on **Replica Set Status** or **Shard Cluster Status** and then on the cluster/set that is to be restored.

    1. On the right-hand side of the screen, click on the **Restore** button.
    2. This should give a list of available snapshots, or you may enter a custom time. Check the **Use Custom Point In Time** checkbox. Click on the **Date** field and select a date and a time to which you want to restore the data to, in hours and minutes, and click on **Next**. Note that the **Point in Time** feature only restores to a point in the last 24 hours.

    Here you would be asked the HTTPS or SCP format. Subsequent steps are similar to what we did on a previous occasion step 14 and 15 onwards.

How it works...

After the backup for a replica set was set up, we first imported some random data into the test database so that we can expect that to be sent to the MMS backup service that we would restore at a later point in time. We saw how to exclude namespaces from being backed up in steps 2, 3, 4, and 5.

Now, looking at the snapshot and retention policy settings, we can see we have the choice of the time interval in which the snapshots are to be taken and the number of days for which they are to be retained (step 9). We can see that, by default, snapshots are taken every 6 hours and they are saved for 2 days. The snapshot that is taken at the end of the day gets saved for a week, which is 7 days. The snapshot taken at the end of the week and month is saved for 4 weeks and 13 months respectively. A snapshot can be taken once every 6, 8, 12, and 24 hours. However, one needs to understand the flip side of taking snapshots after long time durations. Suppose the last snapshot is taken at 18:00 hours, getting the data as of 18:00 hours for restore is very easy, as it is stored on the MMS backup servers. However, we need the data as of 21:30 hours for restoration. As MMS backup supports point-in-time backup, it would use the base snapshot as 18:00 hours and then just replay the changes on it after the snapshot is taken, till 21:30 hours. This replaying is similar to how an oplog would be replayed on the data. There is a cost for this replay and thus, getting point-in-time backup is slightly more expensive than getting the data from a snapshot. Here we had to replay the data for 3.5 hours, from 18:00 hours to 21:30 hours. Imagine if the snapshots were set to be taken after 12 hours and our first snapshot was taken at 00:00 hours; we would have snapshots at 00:00 hours and 12:00 hours every day. To restore the data as of 21:30 hours, with 12:00 hours as the last snapshot, we will have to replay 9.5 hours of data, which is much more expensive. More frequent snapshots means more storage space usage but less time needed to restore a database to a given point in time. At the same time, less frequent snapshots require less storage but at the cost of more time to restore the data to a point in time. You need to decide and have a trade-off between these two, space and time for restoration. For the daily snapshot, we can choose the retention from 3-180 days. Similarly, for the weekly and monthly snapshots, the retention period can be chosen between 1-52 weeks and 1-36 months, respectively.

The screenshot in step 9 has a column for the expiry of the snapshot. For the first snapshot taken it is is 1 year, whereas others expire in 2 days. The expiration is as per what we discussed earlier. On changing the expiration values, the old snapshots are not affected or adjusted as per the changed times. The new snapshots taken will however be as per the modified settings for retention and frequency.

We saw how to download the dump (step 10 onwards) and then use it to restore the data in the database. It was pretty straightforward and doesn't need a lot of explanation, except for a few things. First, if the data is for a shard, there will be multiple folders, one for each shard; and each of them will have the database files as against what we saw here in the case of a replica set, where we have a single folder with database files in it.

Monitoring and Backups

Finally, let us look at the screen when we choose SCP as the option:

	Select Restore Destination ✕
Snapshot / Point In Time	04/15/14 - 11:40
Delivery Method	Push via Secure Copy (SCP) ▼
Format ⓘ	Individual DB Files ▼
SCP Host	
SCP Port	22
SCP User	
Auth Method	Password
Password	[Test]
Target Directory	
	Leave blank for home directory.
	Note - push restores will originate from IP ranges 64.70.114.115/32 or 4.71.186.0/24. However, to use the test button above, you must whitelist all IPs.
Back	Authenticate

SCP is for secure copy. The files will be copied over a secure channel to a machine's filesystem. The host that is given needs to have a public IP, which will be used to SCP the files. This makes a lot of sense when we want the data from MMS to be delivered to a machine running on Unix OS on the cloud, say one of the AWS virtual instances. Rather than getting the file using HTTPS on our local machine and then reuploading it to the server on the cloud, you can specify the location where the data needs to be copied in the **Target Directory** block, the hostname, and the credentials. There are a couple of ways for authentication as well; a password is an easy way with an additional option to SSH key pair. If you have to configure the firewalls of your host on the cloud to allow incoming traffic over the SSH port, the public IP addresses are given at the bottom of the screen (`64.70.114.115/32 or 4.71.186.0/24` in our screenshot), which you should whitelist to allow incoming secure copy request over port 22.

See also

- We have seen running backups using MMS, which uses oplogs for this purpose. The *Implementing triggers in Mongo using oplog* recipe in *Chapter 5, Advanced Operations*, uses oplog to implement trigger-like functionalities. This concept is the backbone of the real-time backup used by the MMS backup service.

7
Cloud Deployment on MongoDB

In this chapter, we will cover the following recipes:

- Setting up and managing the MongoLab account
- Setting up a sandbox MongoDB instance on MongoLab
- Performing operations on MongoDB from the MongoLab GUI
- Setting up MongoDB on Amazon EC2 using the MongoDB AMI
- Setting up MongoDB on Amazon EC2 without using the MongoDB AMI

Introduction

Though explaining cloud computing is not in the scope of this book, I will explain it in just one paragraph. Any business, big or small, needs hardware infrastructure and different software installed on it. An operating system is the basic software needed, along with different servers (from a software perspective) for storage, mail, Web, database, DNS, and so on. The list of software frameworks/platforms needed might end up being large. The point of interest here is that the initial budget for this hardware and software platform is high; we are not even considering the real estate needed to host it. This is where cloud computing providers such as Amazon, Rackspace, Google, and Microsoft come into play. They have hosted high-end hardware and software in different data centers across the globe and let us choose from different configurations to start an instance. Then, this is accessed remotely over the public network for management purposes. Literally, all our setup is done in the cloud provider's data center, and we just pay as we use. Shut down the instance and you stop paying for it. Not only small start-ups but large enterprises also often temporarily fall back to cloud servers for a temporary rise in the computing resource demand. The prices offered by the providers are very competitive too; particularly, **Amazon Web Service** (**AWS**) of all of them in my opinion and its popularity says it all.

Cloud Deployment on MongoDB

The wiki page at `http://en.wikipedia.org/wiki/Cloud_computing` has a lot of detail, perhaps a bit too much for someone new to the concept, but it is a good read, nevertheless. The article at `http://computer.howstuffworks.com/cloud-computing/cloud-computing.htm` is pretty good and I recommended that you read it if you are not aware of the concept of cloud computing.

In this chapter, we will set up MongoDB instances on the cloud using MongoDB service providers and then, by ourselves on AWS.

Setting up and managing the MongoLab account

In this recipe, we will be evaluating one of the vendors, MongoLab, providing MongoDB as a service. This introductory recipe will introduce to you what MongoDB as a service is, and then it will demonstrate how to set up and manage an account in MongoLab (`https://mongolab.com/`).

In all the recipes in this book, we have covered setting up, administering, monitoring, and developing the instances of MongoDB in the organizations/personal premises so far. This not only needs man power with the appropriate skill set to manage the deployments, but also appropriate hardware to install and run Mongo servers. This needs large investments upfront that might not be a viable solution for start-ups or even organizations that are not clear about adopting this technology or migrating to it. They might want to evaluate it and see how it goes before moving full fledged to this solution. What would be ideal is to have a service provider that takes care of hosting the MongoDB deployments, managing, and monitoring the deployments, and providing support. The organizations that opt for these services need not invest upfront in setting up the servers nor recruit or outsource to consultants for the administration and monitoring of the instances. All that one needs to do is choose the hardware and software platform, configuration, and the appropriate MongoDB version, and set up an environment from a user-friendly GUI. It even gives you an option to use your existing cloud provider's servers.

Having explained in brief what these vendor-hosting services do and why they are needed, we will start this recipe by setting up an account with MongoLab and see some basic user and account management. MongoLab is by no means the only hosting provider for MongoDB. You might also want to take a look at `http://www.mongohq.com/` and `http://www.objectrocket.com/`. At the time of writing this book, MongoDB itself started providing MongoDB as a service on Azure cloud and is currently in beta phase.

Chapter 7

How to do it...

1. Visit `https://mongolab.com/signup/` to sign up. If you don't have an account created, just fill in the relevant details and create an account.

2. Once the account is created, click on the **Account** link in the top-right corner of the page, as shown in the following screenshot:

3. Click on the **Account Users** tab in the top-left corner; it should be selected by default:

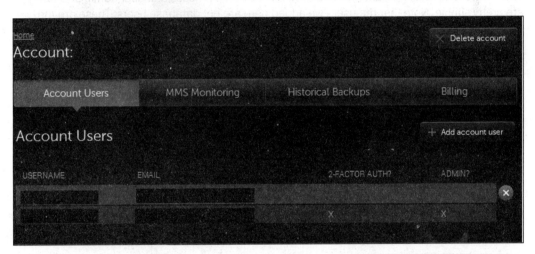

4. To add a new account, click on the **+ Add account user** button. One pop-up window will ask for the username, e-mail ID, and password of the user. Enter the relevant details, and click on the **Add** button.

5. Click on the user, and you will be able to navigate to a page where you can change the username, e-mail ID, and password. You might transfer the admin rights to the user by clicking on the **Change to Admin** button on this page.

6. Similarly, by clicking on your own user details, you will have the options to change the username, e-mail ID, and password.

7. Click on the **Set up two-factor authentication** button to activate the multifactor authentication using Google Authenticator. You need to have the Google Authenticator installed on your Android, iOS, or BlackBerry phone to proceed with the setup of multifactor authentication.

Cloud Deployment on MongoDB

8. On clicking on the button, we should see the QR code that can be scanned using the Google Authenticator, or if scanning is not possible, click on the URL underneath the QR code, which will show the code. Manually set up a time-based account in the Google Authenticator. There are two types of Google Authenticator accounts: time based and counter based. For more details, visit `http://en.wikipedia.org/wiki/Google_Authenticator`.

9. Similarly, you can delete users from the **Account** page by clicking on the cross next to the user's row under **Account Users**.

How it works...

There is nothing much to explain in this section. The setup process and user administration are pretty simple. Note that the users we added here are not database users. These are the users that have access to **Account** on MongoLab, for which we added them. The account can be the name of the organization and can be seen at the top of the screen. The multifactor authentication account set up in the Google Authenticator software on the handheld device should not be deleted, as whenever the user logs in to the MongoLab account from the browser, he will be asked to enter the Google Authenticator account to continue.

Setting up a sandbox MongoDB instance on MongoLab

In the previous recipe, we saw how to set up an account on MongoLab and add users to your account. We still haven't seen how to fire up an instance on the cloud and use it to perform some simple operations. In this recipe, this is exactly what we will do.

Getting ready

Refer to the previous recipe to set up an account with MongoLab. We will set up a free sandbox instance. We will require some way to connect to this started Mongo instance, and thus, we will need a Mongo shell, which comes only with the complete Mongo installation, or you might choose to use a programming language of your choice to connect to the started Mongo instance. Refer to *Chapter 3*, *Programming Language Drivers*, for recipes on connecting and performing operations using a Java or Python client.

How to do it...

1. Go to the home page at `https://mongolab.com/home` and click on the **Create new** button.
2. Select a cloud provider; for this example, we chose **Amazon Web Services**.

Chapter 7

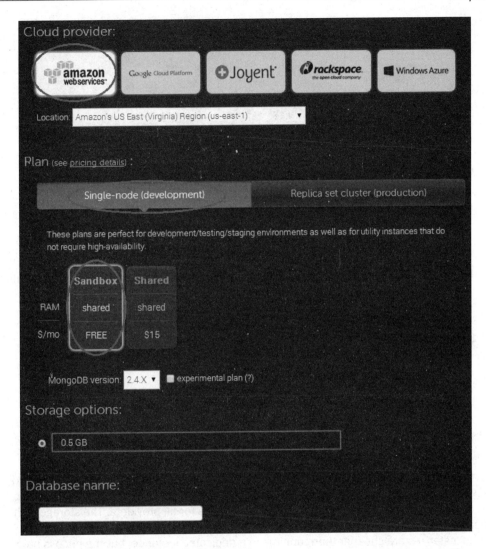

3. Click on **Single-node (development)** and then on the **Sandbox** option. Do not change the location of the cloud server, as the free sandbox instance is not available in all data centers. Since this is sandbox we are ok with any location.

4. Add any name for your database. The name I chose is `mongolab-test`. Click on **Create new MongoDB deployment** after entering the name.

5. This will take you to the home page, and the database will now be visible. Click on the instance name. The page here shows the details of the MongoDB instance selected. The instruction to connect from the shell or programming language is given at the top of the page, along with the public hostname of the started instance.

Cloud Deployment on MongoDB

6. Click on the **Users** tab and then on the **Add database user** button.

7. In the pop-up window, add the username and password as `testUser` and `testUser`, respectively (or any of your choice).

8. With the user added, start the Mongo shell as follows, assuming that the name of the database is `mongolab-test`, and the username and password is `testUser`:

   ```
   $ mongo <host-name>/mongolab-test -u testUser -p testUser
   ```

9. On connecting, execute the following command in the shell and check if the database name is `mongolab-test`:

   ```
   > db
   ```

10. Insert one document in a collection as follows:

    ```
    > db.messages.insert({_id:1, message:'Hello mongolab'})
    ```

11. Query the collection as follows:

    ```
    > db.messages.findOne()
    ```

How it works...

The steps executed are very simple. We created one shared sandbox instance in the cloud. MongoLab itself does not host the instances but uses one of the cloud providers to do the hosting. MongoLab does not support sandbox instances for all providers. The storage with the sandbox instance is 0.5 GB and is shared with other instances on the same machine. Shared instances are cheaper than running on a dedicated instance, but the price is paid in performance. The CPU and I/O are shared with other instances, and thus, the performance of our shared instance is not necessarily in our control. For a production use case, shared instance is not a recommended option. Similarly, we need to set up a replica set when running in production. If we look at the screenshot in step 2, we will see another tab next to the **Single-node (development)** option. This is where you might choose the configuration for the machine in terms of RAM and disk capacity (and the price too) and set up a replica set.

Chapter 7

	Shared	M1	M2	M3	M4	M5	M6	M8
RAM	shared	1.7 GB	3.7 GB	7.5 GB	15 GB	34.2 GB	68.4 GB	244 GB
$/mo	$89	$200	$400	$850	$1600	$2800	$4800	$8600

These plans are suited for production environments that demand high-availability, as they feature multiple nodes with automatic failover. All plans come standard with two data nodes plus one arbiter node; additional nodes available on request.

MongoDB version: 2.4.10 ▼
- 2.0.9
- 2.2.7
- 2.4.10

Storage options:

◉	200GB Standard EBS Volume	included
○	200GB Provisioned IOPS EBS Volume (SSD option)	$450
○	600GB Standard EBS Volume	$300
○	600GB Provisioned IOPS EBS Volume (SSD option)	$850

As you can see, you get to choose which version of MongoDB to use. Even if a new version of MongoDB gets released, MongoLab will not start supporting it immediately, as it usually waits for a few minor versions to be rolled out before supporting the new version for production users. Also, when we choose a configuration, the default available option is two data nodes and one arbiter, which is sufficient for the majority of use cases.

The RAM and disk chosen depend completely on the nature of the data and how query/write intensive it is. This sizing is something we do irrespective of whether we are deploying on our own infrastructure or on the cloud. The working set is something that is important to be known before we choose the RAM of the hardware. POC and experiments are done to deal with a subset of data, and then, the estimation can be done for the entire dataset. Refer to the *Estimating the working set* recipe in *Chapter 4*, *Administration*, to estimate the working set on your sample dataset. If the I/O activity is high and low I/O latency is desired, you might even opt for SSD, as we saw in the preceding screenshot. Standalone instances are as good as replica sets in terms of scalability, but not in terms of availability. Thus, we might choose standalone instances for such estimation and development purposes. Shared instances, both free and paid, are good candidates for development purposes. Note that shared instances cannot be restarted on demand as we can for dedicated instances.

Cloud Deployment on MongoDB

What cloud provider do we choose? If you already have your application servers deployed in the cloud, obviously, it has to be the same vendor as your existing vendor. It is recommended that you use the same cloud vendor for the application server and database. Also, they are both deployed on the same location to minimize latency and improve performance. If you are starting afresh, then invest some time in choosing the cloud provider. Look at all other services that the application will need, such as the storage, compute, and other services including e-mails, notification services, and so on. All this analysis is outside the scope of this book, but once you are done with this and finalized with a provider, you might accordingly choose the provider to use in MongoLab. As far as pricing goes, all the leading providers offer competitive pricing.

Performing operations on MongoDB from MongoLab GUI

In the previous recipe, we saw how to set up a simple sandbox instance for MongoDB in the cloud using MongoLab. In this recipe, we'll build on it and see what services MongoLab provides from the perspectives of management, administration, monitoring, and backup.

Getting ready

Refer to the previous recipe to know how to set up a sandbox instance in the cloud using MongoLab.

How to do it...

1. Go to `https://mongolab.com/home`; you should see a list of databases, servers, and clusters. If you have followed the previous recipe, you would see one standalone database, `mongolab-test` (or whatever name you chose for the database). Click on the database name; this will take you to the database details page.

2. On clicking on the `Collections` tab, which should be selected by default, we will see a list of collections present in the database. If the previous recipe was executed before this one, you would see one collection message in the database.

3. Click on the name of the collection, and we will be navigated to the collection details page as follows:

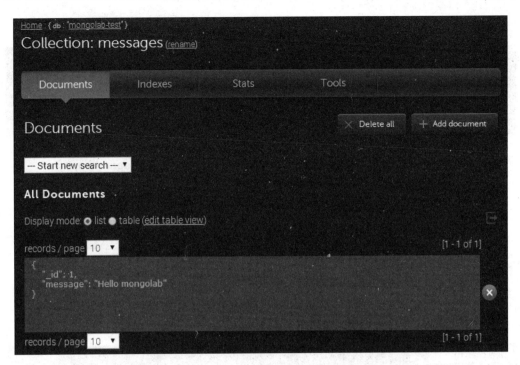

4. Click on the **Stats** option to view the stats of the collection. Except for whether the collection and the maximum number of documents in a collection are capped or not, the contents come as a result of the following command:

   `db.<collectionName>.stats()`

5. In the **Documents** tab, we can query the collection. By default, we will see all the documents with 10 documents shown per page, which can be changed from the **records/page** drop-down list. A maximum value of 100 can be chosen.

6. There is another way to view the documents, which is as a table. Click on the **table** radio button in **Display mode** and click on the **(edit table view)** link to create/edit the table view. In the popup shown, enter the following document for the messages collection and click on **Submit**:

   ```
   {
     "id": "_id",
     "Message Text": "message"
   }
   ```

Cloud Deployment on MongoDB

On doing this, the display will change as follows:

7. From the **—Start new search—** drop-down list, select the **[new search]** option, as shown in the following screenshot:

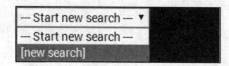

8. With the new query, we will see the following fields to let us enter the query string, sort order, and projections. Enter the query as `{"_id":1}` and fields as `{"message":1, "_id":0}`.

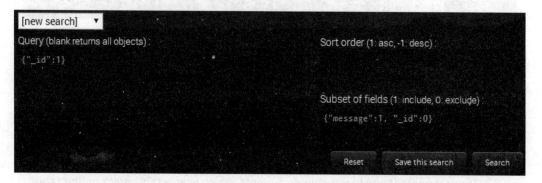

9. You might choose to save the query by clicking on the **Save this search** button and give a name to the query to be saved.
10. Individual documents can be deleted by clicking on the cross next to each record. Similarly, the **Delete all** button will delete all the contents of the collection.
11. Similarly, clicking on **+ Add document** will display an editor to type in the document that will be inserted into the collection. As MongoDB is schemaless, the document need not have a fixed set of fields; the application should make sense out of it.
12. Go to `https://mongolab.com/databases/<your database name>` (`mongolab-test` in this case), which can also be reached by clicking on the database name from the home page.

13. Click on the **Stats** tab next to the **Users** tab. The content shown in the table is the result of the `db.stats()` command.
14. Similarly, click on the **Backups** tab at the top, next to the **Stats** tab. Here, we can select options to take a recurring or one-time backup.
15. When you click on **Schedule recurring backup**, you will get a pop-up window that will let you enter the details of the scheduling, such as the frequency of the backup, the time of the day when the backup needs to be taken, and the number of backups to keep.
16. The backup location can be chosen to be either MongoLab's own **Simple Storage Service (S3)** bucket or the Rackspace cloud file. You might choose to use your own account's storage, in which case, you will have to share the AWS access key/secret key or user ID/API key in case of Rackspace.

How it works...

Steps 1 to 5 are pretty straightforward. In step 6, we provided a JSON document to show the results in a tabular format. The format of the document is as follows:

```
{
  <display column 1> : <name of the field in the JSON document> ,
  <display column 2> : <name of the field in the JSON document> ,

  <display column n> : <name of the field in the JSON document>
}
```

The key is the name of the column to display, and the value is the name of the field in the actual document whose value will be shown as the value of that column. To get a clear understanding, look at the document defined for the messages collection, look at the document in the messages collection, and then take a look at the displayed tabular data. The following is the JSON document we provided; it states the name of the column as the value of the key and the actual field in the document as the value of the column:

```
{
  "id": "_id",
  "Message Text": "message"
}
```

Also, note that the field name and values of the JSON documents here are enclosed in quotes. The Mongo shell is lenient in the sense that it allows us to give field names without quotes.

Cloud Deployment on MongoDB

If we see step 16, we will see that the backups are stored either in MongoLab's AWS S3/Rackspace Cloud Files or in your custom AWS S3 bucket/Rackspace Cloud Files. In latter cases, you need to share your AWS/Rackspace credentials with MongoLab. If this is a concern and the credentials can potentially be used to access other resources, it is recommended that you create a separate account and use it for backup purposes from MongoLab. You might also use the backup created to create a new MongoDB server instance from MongoLab. Needless to say, if you have used your own AWS S3 bucket/Rackspace Cloud Files, storage charges are additional; they are not a part of MongoLab's charges.

There are some important points worth mentioning. MongoLab provides a REST API for various operations. The REST API too can be used in place of the standard drivers to perform CRUD operations. However, using MongoDB client libraries is the recommended approach. One good reason to use the REST API right now over the language driver is if the client is connecting to the MongoDB server over public network. The shell we started on our local machine that connects to the MongoDB server on the cloud sends unencrypted data to the server, which makes it vulnerable. On the other hand, if REST APIs are used, the traffic is sent over a secure channel as HTTPS is used. MongoLab plans to support a secure channel for communication between the client and the server in future, but at the time of writing this book, it is not available. If the application and database are in the same data center of the cloud provider, you are safe and depend on the security provided by the cloud provider for their local network, which generally is not a concern. However, there is nothing you can do for secure communication other than ensuring that your data doesn't go over public networks.

One more scenario where MongoLab doesn't work is when you want the instances to run on your own instance of a virtual machine rather than on the one chosen by MongoLab or when we want the application to be in a virtual private cloud. Cloud providers provide services such as Amazon VPC, where a part of the AWS cloud can be treated as a part of your network. If you intend to deploy your MongoDB instance in such an environment, MongoLab cannot be used.

Setting up MongoDB on Amazon EC2 using the MongoDB AMI

In the earlier few recipes, we saw how to start MongoDB in the cloud using a hosted service provided by MongoLab, which gave an alternative to set up MongoDB on all leading cloud vendors. However, if we plan to host and monitor the instance ourselves for greater control or set up within our own virtual private cloud, we can do it ourselves. Though the procedure varies from cloud provider to cloud provider, we will demonstrate it using AWS. There are a couple of ways to do this, but in this recipe, we will do it using the **Amazon Machine Image** (**AMI**). The AMI is a template that contains details such as the operating system and the software that will be available on the started virtual machine. All this information will be used while booting up a new virtual machine instance on the cloud. To know more about the AMI, visit http://en.wikipedia.org/wiki/Amazon_Machine_Image.

Talking about AWS, **Elastic Cloud Compute** (**EC2**) is a service that lets you create, start, and stop servers of different configurations in the cloud that run on operating systems of your choice (the prices differ accordingly). Similarly, Amazon **Elastic Block Store** (**EBS**) is a service that provides persistent block storage with high availability and low latency. Initially, each instance has a store known as the ephemeral store attached to it. This is a temporary store, and the data might be lost when the instance restarts. EBS block storage is thus attached to the EC2 instance to maintain persistence even when the instance is stopped and then restarted. Standard EBS doesn't promise a minimum guarantee for the **I/O operations per second** (**IOPS**). For moderate workload, the default of about 100 IOPS is ok. However, for high-performance I/O, EBS blocks with guaranteed IOPS are also available. The pricing is more as compared to the standard EBS block, but it is a good option to opt for if low IO rate can be a bottleneck in the performance of the system.

Getting ready

The first thing you need to do is sign up for an AWS account. Visit `http://aws.amazon.com/` and click on **Sign Up**. Log in if you have an Amazon account; otherwise, create a new one. You will have to give your credit card details, although the recipes we have here will use the free micro instance unless we explicitly mention otherwise. We will connect to the instance on the cloud using PuTTY. You can download and install PuTTY on your machine if you have not already done so. It can be downloaded from `http://www.putty.org/`.

For the installation of using AMI, we cannot use the micro instance and will have to use the minimum of standard large. Get more details on the pricing of EC2 instances in different regions at `https://aws.amazon.com/ec2/pricing/`. Choose the appropriate region based on the geographical and financial factors.

1. The first thing you need to do is create a key pair if you have not already created one. Steps 1 to 5 are only for the creation of the key pair. This key pair will be used to log into the Unix instance started in the cloud from the PuTTY client. Skip to step 6 if the key pair is already created and the `.pem` file is available with you.

Cloud Deployment on MongoDB

2. Go to `https://console.aws.amazon.com/ec2/` and make sure the region you have in the top-right corner (as shown in the following screenshot) is the same as the one in which you are planning to set up the instance:

3. Once the region is selected, the page with the **Resources** heading will show all the instances, key pairs, IP addresses, and so on for this region. Click on the **Key Pairs** link; this will navigate you to the page where all the existing key pairs will be shown, and you can create new ones.

4. Click on the **Create Key Pair** button, and in the pop-up window type in any name of your choice. Let's say, we call it `EC2 Test Key Pair`, and click on **Create**.

5. Once the key pair is created, a `.pem` file will be generated. Ensure that the file is saved, as this will be needed for subsequent access to the machine.

6. Next, we will convert this `.pem` file to a `.ppk` file to be used with PuTTY.

7. Start PuTTYgen. If it is not already available, it can be downloaded from `http://www.chiark.greenend.org.uk/~sgtatham/putty/download.html`.

8. We will see the following screenshot:

Chapter 7

9. Select the **SSH-2 RSA** option and click on the **Load** button. In the file, select **All files** and select the `.pem` file that was downloaded when the key pair was generated in the EC2 console.

10. Once the `.pem` file is imported, click on the **Save private key** option and save the file with any name. This time, the file is a `.ppk` file. Save this file to log in to the EC2 instance from PuTTY in future.

How to do it...

1. Visit the Amazon market place at `https://aws.amazon.com/marketplace/` and search for MongoDB, as shown in the following screenshot:

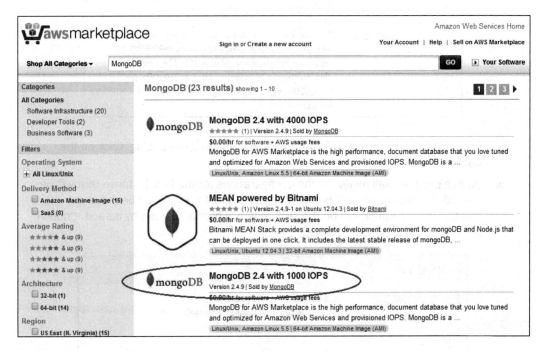

2. Look out for AMIs by MongoDB, as these are the official ones sold by MongoDB. There are different AMIs available with different provisioned I/O rates. For this example, we will choose the one with 1000 IOPS of data.

Cloud Deployment on MongoDB

3. Click on **Name of the Image**, which is a URL, and we will be navigated to the details page. The following portion of the details page is of particular our interest. Notice the **Highlights** section. There are three additional EBS volumes that are reserved and will be attached to this instance: one will be used for data (with the highest IOPS), one for journal, and one for logs (with the lowest IOPS).

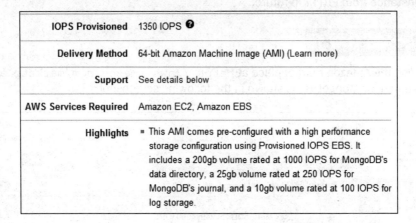

4. The page will also provide information on the AMI and the pricing. Click on the **Continue** button.
5. On this page, we will review all the configurations for the EC2 instance that will be started. The first option will be the MongoDB version, which will be the latest one, and the second option will be the AWS region, which is US East by default. Choose the version and region if required.

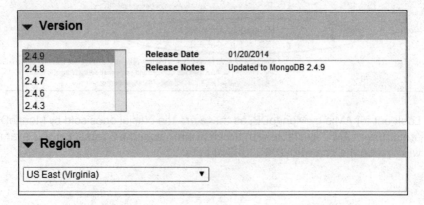

6. The next option is to select the instance. We will choose the **Standard Large (m1. large)** option for our test. Leave the VPC settings set to default.

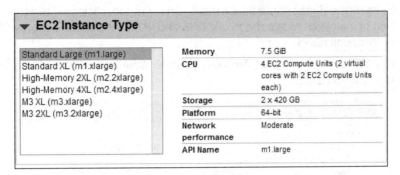

7. The next setting is the **Security** settings that allow connections from the entire World to the started instance of EC2. We will choose to use the settings recommended by the vendor of the AMI. You are free to use any other security policy if you have defined one earlier in EC2.

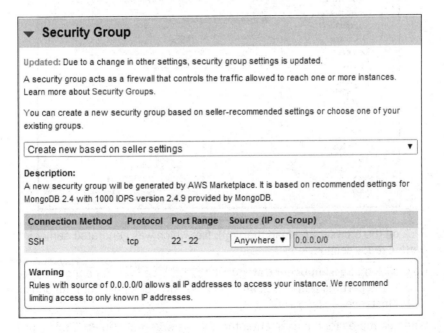

8. Finally, select the key pair, the one we created in the *Getting ready* section. Once done, click on **Accept Terms & Launch with 1-Click**.

Cloud Deployment on MongoDB

9. Visit the EC2 console from the browser and click on **Instances** on the left-hand side menu.

10. The instance will take some time to start. Once started, click on the instance name in the list of instances to see the public DNS and IP address at the bottom of the page. Copy this public DNS.

11. Start PuTTY, and click on the **Auth** option under **Connection/SSH**, as shown in the following screenshot:

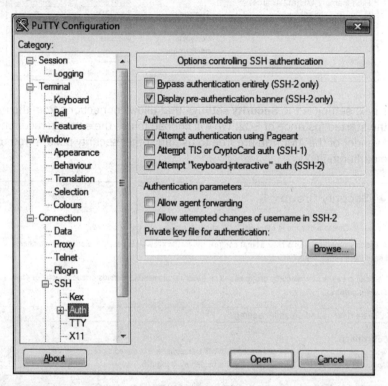

12. Click on the browser and load the .ppk file, which was generated earlier in the *Getting ready* section.

13. Now, click on **Session** under **Category** and enter the hostname that was copied in step 8. The port will remain 22, as this is the only open port from the public network to this instance.

14. When prompted for the user, enter the user as ec2-user in PuTTY. The private key loaded in steps 9 and 10 will be used for authentication, and you do not need to enter a password.

15. We will use /data for the data and /logs to save the logs. These two are configurable parameters. The journal is always created in /data/journal and is not configurable. Refer to step 3 where we mentioned that there are three EBS volumes associated with this EC2 instance.

Chapter 7

16. Execute the following command to start a `mongod` instance with logs written to the `mongo.log` file in the `/logs` directory and the process to run the background:

    ```
    $ sudo mongod --logpath ~/logs/mongo.log --smallfiles --oplogSize 50 --fork
    ```

17. Now, start a Mongo client from the shell as follows:

    ```
    $ mongo
    ```

18. Execute the following command from the Mongo shell after connecting to the Mongo instance:

    ```
    > db.ec2Test.insert({_id:1, msg:'Hello, My first Mongo instance on cloud'})
    > db.ec2Test.find()
    { "_id" : 1, "msg" : "Hello, My first Mongo instance on cloud" }
    >
    ```

    Congratulations! Now, we have successfully started a standalone MongoDB instance on an EC2 instance.

How it works...

In step 6, we saw how to set up the security for the started instance. We configured it to just allow incoming traffic for SSH over port 22 from all hosts from the public network. For tighter security, rather than allowing traffic from all hosts (0.0.0.0), we can allow traffic from a limited set of IP addresses. Let's see if we can connect to the MongoDB instance started over the cloud from the Mongo shell on the local machine. For this activity, we will need MongoDB set up on the local machine; if not, you might just read through the content and understand the concept.

1. Note the public IP address/hostname of the instance started in the cloud and enter the following command on your local machine's command line:

    ```
    $ mongo --host <Public host name of the cloud instance>
    ```

2. We will see that this operation fails with the following exception on the console:

    ```
    MongoDB shell version: 2.4.6
    connecting to: ec2-54-87-4-215.compute-1.amazonaws.com:27017/test
    Sat May 03 14:30:23.376 Error: couldn't connect to server ec2-54-87-4-215.compute-1.amazonaws.com:27017 at src/mongo/shell/mongo.js:147
    exception: connect failed
    ```

    This is simply because the incoming traffic from a public network to this server over port `27017` is blocked. In fact, all traffic, except that on port `22`, is blocked.

Cloud Deployment on MongoDB

3. We will now open port `27017` for our current IP address. Note that this is not a recommended approach in the production environment; we are just doing this to test connecting to the instance on the cloud. Instead, the correct way is to just open the SSH connection to the cloud instance and then connect to the server from a client run over this instance, as we did in the previous section.

4. Go to the EC2 console, choose the correct region at the top, and then click on the **Security Groups** menu option on the left-hand side. We will see the security groups defined as follows:

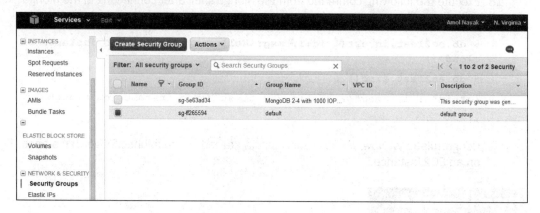

5. As we can see, there is a group that is created when we started the instance. Click on this group to see the details at the bottom of the screen and a way to edit the rules. Select the type as **Custom TCP Rule**, port as **27017**, and source as **My IP/Custom IP**, where you can enter any IP address or all IP addresses. We will choose **My IP** in this case for testing purpose and click on **Save**, as shown in the following screenshot:

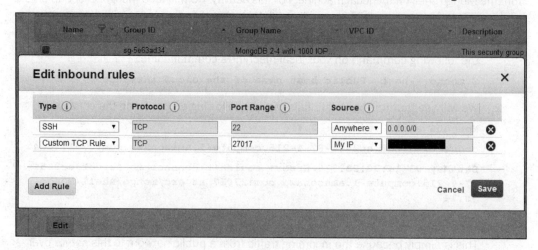

```
$ mongo --host <Public host name of the cloud instance>
```

6. This time, we will be able to connect to this instance. Now, connect to this MongoDB instance started in the cloud by typing in the following command from your local machine's operating system shell:

```
$ mongo --host <Public host name of the cloud instance>
```

What we saw was a very simple demo of what a security group of this cloud instance does. For more details on EC2 instance security, visit http://docs.aws.amazon.com/AWSEC2/latest/UserGuide/EC2_Network_and_Security.html.

What we saw so far was how to start the mongod instance on the cloud using the MongoDB AMI. If this is your only objective, then the rest of the contents in the section can be skipped. What we will see now is how the filesystem is set up on this instance.

Moving on, let's look at the filesystem setup. From the shell of the started MongoDB instance, execute the following command:

```
[ec2-user@ip-10-236-144-125 ~]$ mount
/dev/xvda1 on / type ext4 (rw,noatime)
proc on /proc type proc (rw)
sysfs on /sys type sysfs (rw)
devpts on /dev/pts type devpts (rw,gid=5,mode=620)
tmpfs on /dev/shm type tmpfs (rw)
/dev/xvdf on /data type ext4 (rw,noexec,noatime)
/dev/xvdg on /journal type ext4 (rw,noexec,noatime)
/dev/xvdh on /log type ext4 (rw,noexec,noatime)
none on /proc/sys/fs/binfmt_misc type binfmt_misc (rw)
```

We can see that there are three different mount points for /data, /journal, and /log. These are the three provisioned EBS storage blocks with 1000, 250, and 100 IOPS, respectively. The journal is, however, created in the journal directory in the data directory. Let's list the files in the /data directory as follows:

```
[ec2-user@ip-10-236-144-125 ~]$ ls -al /data
total 98344
drwxr-xr-x   4 mongod mongod    4096 May  3 08:35 .
dr-xr-xr-x  26 root   root      4096 May  3 08:23 ..
lrwxrwxrwx   1 root   root         8 Apr 28  2013 journal -> /journal
```

As we can see, in the partial captured output of the ls command, the journal directory in the /data directory is a link to the /journal mount.

Finally, let's execute the following command:

```
[ec2-user@ip-10-236-144-125 ~]$ cat /etc/fstab
#
LABEL=/         /               ext4    defaults,noatime  1  1
tmpfs           /dev/shm        tmpfs   defaults          0  0
devpts          /dev/pts        devpts  gid=5,mode=620    0  0
sysfs           /sys            sysfs   defaults          0  0
proc            /proc           proc    defaults          0  0
/dev/sdf /data ext4 defaults,auto,noatime,noexec 0 0
/dev/sdg /journal ext4 defaults,auto,noatime,noexec 0 0
/dev/sdh /log ext4 defaults,auto,noatime,noexec 0 0
```

We will see that the three mount points are already defined in this file. As we use AMIs to create the EBS machine, we get all these things configured.

Let's look at one entry, `/dev/sdf /data ext4 defaults,auto,noatime,noexec 0 0`, added in the file and analyze it's fields one by one. These values are tab separated.

- The first value, `/dev/sdf`, is the device that we are looking to mount
- The second value, `/data`, is the directory to which the directory will be mounted to
- The third parameter, `ext4`, is the type of the filesystem
- Next, we have comma-separated values of options:
    - The value, `default`, is used to load default options for the `ext4` partition.
    - The value, `auto`, is used to indicate that the device will be mounted automatically on startup; `auto` is the default value and need not be explicitly mentioned.
    - Whenever a file is accessed, even in case of read, the last-accessed time of the file on the filesystem is updated by Unix. This will have heavy, negative performance impact on both read and write operations. Setting `noatime` instructs OS to not update this last-accessed time.
    - The `noexec` value instructs that these filesystems cannot have executables on them.
- The final two values are 0 and 0 for dump frequency and pass number. By setting the pass number to 0, we disable partition checks for these partitions

That is pretty much it; as we saw, the AMI has made life easy for us and given a machine image with all the recommended settings to help us get up to speed in spinning off a server in the cloud and starting the MongoDB server. All other steps to start the servers, form replica sets and shards, and monitor them are the same as the steps used to start a server on your local machine or in your own data centers. Refer to *Chapter 4*, *Administration*, and *Chapter 6*, *Monitoring and Backups*, for more recipes on administration and monitoring Mongo instances.

Chapter 7

Make sure you stop the EC2 instance, if this is a test, as soon as possible from the EC2 console to avoid being charged unnecessarily. A stopped instance will not attract any charges. The blocked EBS instances are also charged for data on it; if you plan to not use this instance anymore, terminate the instance and release the EBS volumes attached.

Setting up MongoDB on Amazon EC2 without using the MongoDB AMI

In the previous recipe, we saw how to start a standalone database in the cloud using the MongoDB AMI; this is perhaps the simplest way to start the server on the EC2 instance. However, when using an AMI, you are tied to the configurations the AMI supports. For instance, for a noncrucial instance, you might not want a large instance or an EBS volume with guaranteed IOPS. You might even be OK with the same EBS volume for data, journal, and logs to cut on the costs, as the instance you are setting up is a development or test instance. Also, the operating system for the AMIs is Amazon Linux; if you wish to use a different OS and install MongoDB on it, AMI isn't of any help. In all such scenarios, setting up the instance manually is the only option left. This isn't a simple job, and a careful setup of various factors, including the standard operating system parameters, is needed. In this recipe, we will not get into these complicated tasks but rather set up a small micro instance, which is as good as a sandbox instance with one EBS block volume attached to it.

Getting ready

Refer to the *Getting ready* section of the previous recipe, which is a prerequisite for this recipe as well.

How to do it...

1. Go to https://console.aws.amazon.com/ec2/, click on the **Instances** option in the left-hand corner, and then the **Launch Instance** button.

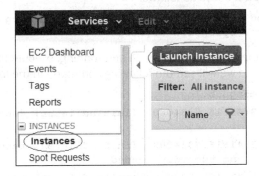

2. As we will start a free micro instance, check the **Free tier only** checkbox on the left-hand side. On the right-hand side, select the instance we want to set up. We will choose to use **Ubuntu Server**. Click on **Select** to navigate to the next window.

3. Choose **Micro Instance** and click on **Review and Launch**. Ignore the security warning; the default security group that you have is the one that will accept connections over port 22 from all the hosts on the public network.

4. Without editing any default settings, click on **Launch**. On clicking **Launch**, a popup will appear, letting you choose an existing key pair. If you proceed without a key pair, you would need the password or have to create a new key pair. In the previous recipe, we already created a key pair; we will use this here.

5. Click on **Launch Instance** to start the new micro instance.

6. Refer to steps 9 to 12 in the previous recipe to learn how to connect to the started instance using PuTTY. Note that we will use the `ubuntu` user instead of `ec2-user`, which we used in the last recipe, as this time we are using Ubuntu instead of Amazon Linux.

7. Before we add a MongoDB repository, we need to import the MongoDB public key as follows:

```
$ sudo apt-key adv --keyserver hkp://keyserver.ubuntu.com:80 --recv 7F0CEB10
```

8. Execute the following command on the operating system shell:

```
$ echo 'deb http://downloads-distro.mongodb.org/repo/ubuntu-upstart/ dist 10gen' | sudo tee /etc/apt/sources.list.d/mongodb.list
```

9. Load the `local` database by executing the following command:

```
$ sudo apt-get install mongodb-org
```

10. Execute the following command to create the required directories:

```
$ sudo mkdir /data /log
```

11. Start the `mongod` process as follows:

```
$ sudo mongod --dbpath /data --logpath /log/mongodb.log --smallfiles --oplogsize 50 -fork
```

12. To ensure that the server process is up and running, execute the following command from the shell, and we will see the following command in the log:

```
$ tail /log/mongodb.log
2014-05-04T13:41:16.533+0000 [initandlisten] journal dir=/data/journal
2014-05-04T13:41:16.534+0000 [initandlisten] recover : no journal files present, no recovery needed
2014-05-04T13:41:16.628+0000 [initandlisten] waiting for connections on port 27017
```

Chapter 7

13. Start the Mongo shell as follows and execute the following command:

```
$ mongo
> db.ec2Test.insert({_id: 1, message: 'Hello World !'})
> db.ec2Test.findOne()
```

How it works...

A lot of steps are self-explanatory. It is recommended that you at least go through the previous recipe as a lot of concepts that are explained there apply for this recipe. A few things that are different are explained in this section. For installation, we chose Ubuntu against Amazon Linux, which is standard when you set up the server using the AMI. Different operating systems have different steps for installation. Visit http://docs.mongodb.org/manual/installation/ for steps to install MongoDB on different platforms. Steps 7 to 9 in this recipe are specific for the installation of MongoDB on Ubuntu. Refer to the https://help.ubuntu.com/12.04/serverguide/apt-get.html page for more details on the apt-get command that we executed here to install MongoDB.

In our case, we chose to have the data, journal, and log folders on the same EBS volume. This is because what we set up is a dev instance. In the case of a prod instance, there would be different EBS volumes with provisioned IOPS for optimum performance. This setup allows us to gain advantage of the fact that these different volumes have different controllers, and thus, concurrent write operations are possible. EBS volumes with provisioned volumes are backed by SSD drives. The production deployment notes at http://docs.mongodb.org/manual/administration/production-notes/ state that MongoDB deployment should be backed by RAID-10 disks. When deploying on AWS, prefer PIOPS over RAID-10. For instance, if 4000 IOPS is desired, then choose the EBS volume with 4000 IOPS rather than a RAID-10 setup with 2 X 2000 IOPS or a 4 X 1000 IOPS setup. This not only eliminates unnecessary complexity but also makes it possible to snapshot a single disk as against dealing with multiple disks in the RAID-10 setup. Speaking of snapshotting, journal log and data are written to separate volumes in the majority of production deployments. This is the scenario where snapshotting doesn't work. We need to flush the DB writes, lock the data for further writes until backup completes, and then release the lock. Visit http://docs.mongodb.org/manual/tutorial/backup-with-filesystem-snapshots/ for more details on snapshotting and backups.

Visit http://docs.mongodb.org/ecosystem/platforms/ for more details on deployment on different cloud providers. There is a section specifically for backups on Amazon EC2 instances. Prefer using AMIs to set up MongoDB instances for production deployments, as demonstrated in the previous recipe, over manually setting up the instances. Manual setup is ok for a small dev purpose where a large instance with EBS volumes with provisioned IOPS is overkill.

See also

- Cloud formation is a way in which you can define templates and automate your instance creation for EC2 instances. Know what cloud formation is at `https://aws.amazon.com/cloudformation/` and `https://mongodb-documentation.readthedocs.org/en/latest/ecosystem/tutorial/automate-deployment-with-cloudformation.html`.

- Visit `http://en.wikipedia.org/wiki/Standard_RAID_levels` and `http://en.wikipedia.org/wiki/Nested_RAID_levels` to know more about RAID.

8
Integration with Hadoop

In this chapter, we will cover the following recipes:

- Executing our first sample MapReduce job using the mongo-hadoop connector
- Writing our first Hadoop MapReduce job
- Running MapReduce jobs on Hadoop using streaming
- Running a MapReduce job on Amazon EMR

Introduction

Hadoop is a well-known open source software for processing large datasets. It also has an API for the MapReduce programming model, which is widely used. Nearly all Big Data solutions have some sort of support to integrate with Hadoop to use its MapReduce framework. MongoDB too has a connector that integrates with Hadoop; it lets us write MapReduce jobs using the Hadoop MapReduce API, process data that resides in the MongoDB/MongoDB dumps, and write the result back to the MongoDB/MongoDB dump files. In this chapter, we will look at some recipes that deal with basic MongoDB and Hadoop integration.

Integration with Hadoop

Executing our first sample MapReduce job using the mongo-hadoop connector

In this recipe, we will see how to build the Mongo Hadoop connector from source and set up Hadoop just for the purpose of running examples in the standalone mode. The connector is the backbone that runs MapReduce jobs on Hadoop using the data in Mongo.

Getting ready

There are various distributions of Hadoop; however, we will use Apache Hadoop (http://hadoop.apache.org/). The installation will be done on a Linux-flavored OS, and I am using Ubuntu Linux. For production, Apache Hadoop always runs on a Linux environment; Windows is not tested for production systems. For development purposes, however, Windows can be used. If you are a Windows user, I would recommend that you install a virtualization environment such as VirtualBox (https://www.virtualbox.org/), set up a Linux environment, and then install Hadoop on it. Setting up VirtualBox and then setting up Linux on it is not shown in this recipe, but this is not a tedious task. The prerequisite for this recipe is a machine with the Linux operating system on it and an Internet connection. The version we set up here is 2.4.0 of Apache Hadoop. At the time of writing this book, the latest version of Apache Hadoop and that supported by the mongo-hadoop connector is 2.4.0.

A Git client is needed to clone the repository of the mongo-hadoop connector to a local filesystem. Refer to http://git-scm.com/book/en/Getting-Started-Installing-Git to install Git.

You will also need MongoDB to be installed on your operating system. Refer to http://docs.mongodb.org/manual/installation/ and install it accordingly. Start the mongod instance that listens to port 27017. You are not expected to be an expert in Hadoop, but some familiarity with it will be helpful. Knowing the concept of MapReduce is important, and knowing the Hadoop MapReduce API will be an advantage. In this recipe, we will explain what is needed to get the work done. You might prefer to get more details on Hadoop and its MapReduce API from other sources. The wiki page at http://en.wikipedia.org/wiki/MapReduce gives enough information on the MapReduce programming.

How to do it...

1. First, install Java, Hadoop, and the required packages.
2. Start by installing JDK on the operating system. Type the following command in the command prompt of the operating system:

   ```
   $ javac -version
   ```

3. If the program doesn't execute and instead, you are told about various packages that contain the `javac` program, we would need to install it as follows:

   ```
   $ sudo apt-get install default-jdk
   ```

   This is all we need to do to install Java

4. Now, download the current version of Hadoop from http://www.apache.org/dyn/closer.cgi/hadoop/common/ and download version 2.4 (or the latest `mongo-hadoop` connector supports).

5. After the `.tar.gz` file is downloaded, execute the following commands in the command prompt:

   ```
   $ tar -xvzf <name of the downloaded .tar.gz file>
   $ cd <extracted directory>
   ```

6. Open the `etc/hadoop/hadoop-env.sh` file and replace `export JAVA_HOME = ${JAVA_HOME}` with `export JAVA_HOME = /usr/lib/jvm/default-java`.

7. We will now get the `mongo-hadoop` connector code from GitHub on our local filesystem. Note that you don't need a GitHub account to clone a repository. Clone the Git project from the operating system's command prompt as follows:

   ```
   $ git clone https://github.com/mongodb/mongo-hadoop.git
   $ cd mongo-hadoop
   ```

8. Create a soft link as follows; the Hadoop installation directory is the same as the one we extracted in step 5:

   ```
   $ ln -s <hadoop installation directory> ~/hadoop-binaries
   ```

   For example, if Hadoop in extracted/installed in the `home` directory, the following command would need to be executed:

   ```
   $ ln -s ~/hadoop-2.4.0 ~/hadoop-binaries
   ```

   By default, the `mongo-hadoop` connector will look for a Hadoop distribution under the `~/hadoop-binaries` folder. So, even if the Hadoop archive is extracted elsewhere, we can create a soft link to it. Once the preceding link is created, we will have the Hadoop binaries in the `~/hadoop-binaries/hadoop-2.4.0/bin` path.

9. We will now build the `mongo-hadoop` connector from source for Apache Hadoop Version 2.4.0 as follows. The build, by default, builds the connector for the latest version; so, as of now, the `-Phadoop_version` parameter can be left out, as 2.4 is the latest anyway:

   ```
   $ ./gradlew jar -Phadoop_version='2.4'
   ```

   This build process will take some time to complete

10. Once the build completes successfully, we are ready to execute our first MapReduce job. We would be doing it using a `treasuryYield` sample provided with the `mongo-hadoop` connector project. The first activity is to import the data to a collection in Mongo.

11. Assuming that the `mongod` instance is up and running and listening to port `27017` for connections and the current directory is the root of the `mongo-hadoop` connector code base, execute the following command:

    ```
    $ mongoimport -c yield_historical.in -d mongo_hadoop --drop examples/treasury_yield/src/main/resources/yield_historical_in.json
    ```

12. Once the import action is successful, we are left with copying two JAR files to the `lib` directory. Execute the following commands from the operating system shell:

    ```
    $ wget http://repo1.maven.org/maven2/org/mongodb/mongo-java-driver/2.12.0/mongo-java-driver-2.12.0.jar
    ```

    ```
    $ cp core/build/libs/mongo-hadoop-core-1.2.1-SNAPSHOT-hadoop_2.4.jar ~/hadoop-binaries/hadoop-2.4.0/lib/
    ```

    ```
    $ mv mongo-java-driver-2.12.0.jar ~/hadoop-binaries/hadoop-2.4.0/lib
    ```

13. The JAR file built for the `mongo-hadoop` core to be copied was named as `mongo-hadoop-core-1.2.1-SNAPSHOT-hadoop_2.4.jar` for the trunk version of the code and built for Hadoop 2.4.0. Change the name of the JAR accordingly when you build it yourself for a different version of the connector and Hadoop. The Mongo driver can be the latest version. Version 2.12.0 is the latest one at the time of writing this book.

14. Now, execute the following command in the command prompt of the operating system shell:

    ```
    ~/hadoop-binaries/hadoop-2.4.0/bin/hadoop jar examples/treasury_yield/build/libs/treasury_yield-1.2.1-SNAPSHOT-hadoop_2.4.jar\
    ```

    `com.mongodb.hadoop.examples.treasury.TreasuryYieldXMLConfig\`

    `-Dmongo.input.split_size=8  -Dmongo.job.verbose=true\`

    `-Dmongo.input.uri=mongodb://localhost:27017/mongo_hadoop.yield_historical.in\`

    `-Dmongo.output.uri=mongodb://localhost:27017/mongo_hadoop.yield_historical.out`

15. The output should print out a lot of things. However, the following line in the output will tell us that the MapReduce job is successful:

    ```
    14/05/11 21:38:54 INFO mapreduce.Job: Job job_local1226390512_0001 completed successfully
    ```

16. Connect the `mongod` instance that runs on a localhost from the Mongo client and execute a `find` query on the following collection:

```
$ mongo
> use mongo_hadoop
switched to db mongo_hadoop
> db.yield_historical.out.find()
```

How it works...

Installing Hadoop is not a trivial task, and we don't need to get into this to try our samples for the `mongo-hadoop` connector. To learn about Hadoop, there are dedicated books and articles available. For the purpose of this chapter, we will simply download the archive and extract and run the MapReduce jobs in the standalone mode. This is the quickest way to get going with Hadoop. All the steps up to step 6 are needed to install Hadoop. In the next couple of steps, we simple cloned the `mongo-hadoop` connector repository. You might also download a stable build version for your version of Hadoop from https://github.com/mongodb/mongo-hadoop/releases if you prefer to not build from source and download directly. We then built the connector for our version of Hadoop (2.4.0) until step 13. From step 14 onwards, we ran the actual MapReduce job to work on the data in MongoDB. We imported the data into the `yield_historical.in` collection, which will be used as an input to the MapReduce job. Go ahead and query the collection from the Mongo shell using the `mongo_hadoop` database to see a document. Don't worry if you don't understand the contents; you want to see in this example what you intend to do with this data.

The next step was to invoke the MapReduce operation on the data. The command Hadoop was executed giving one of JAR's path (`examples/treasury_yield/build/libs/treasury_yield-1.2.1-SNAPSHOT-hadoop_2.4.jar`). This is the JAR file that contains the classes that implement a sample MapReduce operation for `treasury yield`. The `com.mongodb.hadoop.examples.treasury.TreasuryYieldXMLConfig` class in this JAR file is the bootstrap class that contains the `main` method. We will see this class soon. There are lots of configurations supported by the connector. A complete list of configurations can be found at https://github.com/mongodb/mongo-hadoop/blob/master/CONFIG.md. For now, we will just remember that `mongo.input.uri` and `mongo.output.uri` are the collections for input and output, respectively, of the MapReduce operations.

With the project cloned, you might import it into any Java IDE of your choice. We are particularly interested in the project at `/examples/treasury_yield` and the `core` project present in the root of the cloned repository.

Integration with Hadoop

Let's look at the `com.mongodb.hadoop.examples.treasury.TreasuryYieldXMLConfig` class. This is the entry point into the MapReduce method and has a `main` method in it. To write MapReduce jobs for Mongo using the `mongo-hadoop` connector, the main class always has to extend from `com.mongodb.hadoop.util.MongoTool`. This class implements the `org.apache.hadoop.Tool` interface, which has the `run` method and is implemented for us by the `MongoTool` class. All that the `main` method needs to do is execute this class using the `org.apache.hadoop.util.ToolRunner` class by invoking its static `run` method, passing the instance of our main class (an instance of `Tool`).

There is a static block that loads some configurations from two XML files: `hadoop-local.xml` and `mongo-defaults.xml`. The format of these files (or any XML file) is as follows. The root node of the file is the configuration node and multiple property nodes under it:

```
<configuration>
  <property>
    <name>{property name}</name>
    <value>{property value}</value>
  </property>
  ...
</configuration>
```

The `property` values that make sense in this context are all those we mentioned in the URL provided earlier. We instantiate `com.mongodb.hadoop.MongoConfig` wrapping an instance of `org.apache.hadoop.conf.Configuration` in the constructor of the bootstrap class `TreasuryYieldXmlConfig`. The `MongoConfig` class provides sensible defaults, which is enough to satisfy the majority of the use cases. Some of the most important things we need to set in the `MongoConfig` instance are the output and the input formats, the `mapper` and the `reducer` classes, the output key and the value of `mapper`, and the output key and the value of `reducer`. The input and output formats will always be the `com.mongodb.hadoop.MongoInputFormat` and `com.mongodb.hadoop.MongoOutputFormat` classes, respectively; they are provided by the `mongo-hadoop` connector library. For the mapper and reducer output key and the value, we have any of the `org.apache.hadoop.io.Writable` implementation. Refer to the Hadoop documentation for different types of `Writable` implementations in the `org.apache.hadoop.io` package. Apart from these, the `mongo-hadoop` connector also provides us with some implementations in the `com.mongodb.hadoop.io` package. For the `treasury yield` example, we used the `BSONWritable` instance. These configurable values can either be provided in the XML file we saw earlier or can be programmatically set. Finally, we have the option to provide them as `vm` arguments, as we did for `mongo.input.uri` and `mongo.output.uri`. These parameters can be provided either in XML or invoked directly from the code in the `MongoConfig` instance; the two methods are `setInputURI` and `setOutputURI`.

We will now look at the `mapper` and `reducer` class implementations. Here, we will copy the important portion of the class to analyze it. Refer to the cloned project for the entire implementation:

```
public class TreasuryYieldMapper
    extends Mapper<Object, BSONObject, IntWritable, DoubleWritable> {

    @Override
    public void map(final Object pKey,
    final BSONObject pValue,
    final Context pContext)
    throws IOException, InterruptedException {
       final int year = ((Date) pValue.get("_id")).getYear() + 1900;
       double bid10Year = ((Number) pValue.get("bc10Year")).doubleValue();
       pContext.write(new IntWritable(year), new DoubleWritable(bid10Year));
    }
}
```

Our mapper extends the `org.apache.hadoop.mapreduce.Mapper` class. The four generic parameters are for the key class, type of the input value, type of the output key, and the output value, respectively. The body of the `map` method reads the `_id` value from the input document, which is `date`, and extracts the year out of it. Then, it gets the double value from the document for the `bc10Year` field and simply writes to the context key-value pair where the key is the year and the value is the double. The implementation here doesn't rely on the value of the `pKey` parameter passed; this can be used as the key, instead of hardcoding the `_id` value in the implementation. This value is basically the same field that will be set using the `mongo.input.key` property in the XML or using the `MongoConfig.setInputKey` method. If none is set, `_id` is anyway the default value.

Let's look at the reducer implementation (with the logging statements removed):

```
public class TreasuryYieldReducer
extends Reducer<IntWritable, DoubleWritable, IntWritable, BSONWritable> {

    @Override
    public void reduce(final IntWritable pKey, final Iterable<DoubleWritable> pValues, final Context pContext)
        throws IOException, InterruptedException {
       int count = 0;
       double sum = 0;
       for (final DoubleWritable value : pValues) {
          sum += value.get();
```

```
            count++;
        }
        final double avg = sum / count;
        BasicBSONObject output = new BasicBSONObject();
        output.put("count", count);
        output.put("avg", avg);
        output.put("sum", sum);
        pContext.write(pKey, new BSONWritable(output));
    }
}
```

This class extended from `org.apache.hadoop.mapreduce.Reducer` and had four generic parameters again for the input key, input value, output key, and the output value respectively. The input to `reducer` is the output from `mapper`. Thus, if you notice carefully, the type of the first two generic parameters is the same as the last two generic parameters of `mapper` we saw earlier. The third and fourth parameters in this case are the type of the key and the value emitted from `reduce`, respectively. The type of the value is `BSONDocument`, and thus, we have `BSONWritable` as the type.

We now have the `reduce` method that has two parameters: the first one is the key, which is the same as the key emitted from the `map` function, and the second parameter is `java.lang.Iterable` of the values emitted for the same key. This is how standard MapReduce functions work. For instance, if the `map` function gave the key-value pairs as (1950, 10), (1960, 20), (1950, 20), (1950, 30), then `reduce` would be invoked with two unique keys, 1950 and 1960. The value for the key 1950 will be an `Iterable` with (10, 20, 30), whereas that of 1960 will be an `Iterable` of a single element (20). The `reduce` function of the reducer class simply iterates through this `Iterable` of doubles, finds the sum and count of these numbers, and writes one key-value pair where the key is the same as the incoming key and the out value is `BasicBSONObject`, with the sum, count, and average in it for the computed values.

There are some good samples, including the `enron` dataset, in the examples of the cloned `mongo-hadoop` connector. If you would like to play around a bit, I would recommend that you take a look at these example projects too and run them.

There's more...

What we saw here was a ready-made sample that we executed. There is nothing like writing one MapReduce job ourselves to clarify our understanding. In the next recipe, we will write one sample MapReduce job using the Hadoop API in Java and see it in action.

See also

- `http://www.mail-archive.com/hadoop-user@lucene.apache.org/msg00378.html` to know what the `Writable` interface is all about and why you should not use plain old serialization

Writing our first Hadoop MapReduce job

In this recipe, we will write our first MapReduce job using the Hadoop MapReduce API and run it using the `mongo-hadoop` connector that gets the data from MongoDB. Refer to the *MapReduce in Mongo using a Java client* recipe in *Chapter 3, Programming Language Drivers*, to see how MapReduce is implemented using a Java client, how to create test data and problem statements.

Getting ready

Refer to the previous recipe to set up the `mongo-hadoop` connector. The prerequisites of the *Executing our first sample MapReduce job using the mongo-hadoop connector* recipe (which is present in this chapter) and the *MapReduce in Mongo using a Java client* recipe in *Chapter 3, Programming Language Drivers*, are all we need for this recipe. This is a Maven project; thus, Maven needs to be set up and installed. Refer to the *Connecting to a single node from a Java client* recipe in *Chapter 1, Installing and Starting the MongoDB Server*, where we gave the steps to set up Maven in Windows. However, this project is built on Ubuntu Linux, and the following is the command you need to execute from the operating system shell to get Maven:

```
$ sudo apt-get install maven
```

How to do it...

1. We have a Java `mongo-hadoop-mapreduce-test` project, which can be downloaded from the book's website. The project is targeted at achieving the same use case that we achieved in the recipes in *Chapter 3, Programming Language Drivers*, where we used MongoDB's MapReduce framework. We had invoked that MapReduce job using the Python and Java clients on earlier occasions.

2. In the command prompt, with the current directory in the root of the project where the `pom.xml` file is present, execute the following command:

   ```
   $ mvn clean package
   ```

3. The JAR `mongo-hadoop-mapreduce-test-1.0.jar` file will be built and kept in the target directory.

4. With the assumption that the CSV file is already imported into the `postalCodes` collection, execute the following command with the current directory still in the root of the `mongo-hadoop-mapreduce-test` project we just built:

   ```
   ~/hadoop-binaries/hadoop-2.4.0/bin/hadoop \
     jar target/mongo-hadoop-mapreduce-test-1.0.jar \
     com.packtpub.mongo.cookbook.TopStateMapReduceEntrypoint \
     -Dmongo.input.split_size=8 \
   ```

```
        -Dmongo.job.verbose=true \
        -Dmongo.input.uri=mongodb://localhost:27017/test.postalCodes \
        -Dmongo.output.uri=mongodb://localhost:27017/test.
        postalCodesHadoopmrOut
```

5. Once the MapReduce job completes, open the Mongo shell by typing the following command in the operating system command prompt and execute the following query from the shell:

   ```
   $ mongo
   > db.postalCodesHadoopmrOut.find().sort({count:-1}).limit(5)
   ```

6. Compare the output with the ones we got earlier when we executed the MapReduce jobs using Mongo's MapReduce framework (*Chapter 3, Programming Language Drivers*).

How it works...

We have kept the classes very simple and with the fewest possible requirements. We just have three classes in our project: `TopStateMapReduceEntrypoint`, `TopStateReducer`, and `TopStatesMapper`. All these classes are in the same package called `com.packtpub.mongo.cookbook`. The map function of the `mapper` class just writes a key-value pair to the context; here, the key is the name of the state, and the value is an integer value 1. The following line of code is from the `Mapper` function:

```
context.write(new Text((String)value.get("state")), new IntWritable(1));
```

What the `reducer` gets is the same key that is a list of states and an `Iterable` of integer value 1. All that we do is write to the context the same name of the state and the sum of the iterables. Now, since there is no size method in the `Iterable` that can give the count in constant time, we are left with adding up the ones we get in linear time. The following is the code snippet in the `Reducer` method:

```
int sum = 0;
for(IntWritable value : values) {
   sum += value.get();
}
BSONObject object = new BasicBSONObject();
object.put("count", sum);
context.write(text, new BSONWritable(object));
```

We write to context the text string that is the key and the value that is a JSON document that contains the count. The `mongo-hadoop` connector is then responsible for writing to the output collection we have, that is, `postalCodesHadoopmrOut`. The document has the `_id` field whose value is same as the key emitted from the mapper. Thus, when we execute the following query, we will get the top five states with the greatest number of cities in our database:

```
> db.postalCodesHadoopmrOut.find().sort({count:-1}).limit(5)
{ "_id" : "Maharashtra", "count" : 6446 }
{ "_id" : "Kerala", "count" : 4684 }
{ "_id" : "Tamil Nadu", "count" : 3784 }
{ "_id" : "Andhra Pradesh", "count" : 3550 }
{ "_id" : "Karnataka", "count" : 3204 }
```

Finally, the `main` method of the main entry point class is as follows:

```
Configuration conf = new Configuration();
MongoConfig config = new MongoConfig(conf);
config.setInputFormat(MongoInputFormat.class);
config.setMapperOutputKey(Text.class);
config.setMapperOutputValue(IntWritable.class);
config.setMapper(TopStatesMapper.class);
config.setOutputFormat(MongoOutputFormat.class);
config.setOutputKey(Text.class);
config.setOutputValue(BSONWritable.class);
config.setReducer(TopStateReducer.class);
ToolRunner.run(conf, new TopStateMapReduceEntrypoint(), args);
```

All that we do is wrap the `org.apache.hadoop.conf.Configuration` object with the `com.mongodb.hadoop.MongoConfig` instance to set various properties and then submit the MapReduce job for execution using `ToolRunner`.

There's more...

In this recipe, we executed a simple MapReduce job on Hadoop using the Hadoop API, sourcing the data from MongoDB, and writing it to the MongoDB collection. What if we want to write the `map` and `reduce` functions in a different language? Fortunately, this is possible using a concept called Hadoop streaming, where stdout is used as a means to communicate between the program and the Hadoop MapReduce framework. In the next recipe, we will demonstrate how to use Python to implement the same use case as the one in this recipe using Hadoop streaming.

Integration with Hadoop

Running MapReduce jobs on Hadoop using streaming

In the previous recipe, we implemented a simple MapReduce job using the Java API of Hadoop. The use case was the same as the one in the recipes of *Chapter 3, Programming Language Drivers*, where we saw MapReduce implemented using Mongo client APIs in Python and Java. In this recipe, we will use Hadoop streaming to implement MapReduce jobs.

The concept of streaming works based on communication using stdin and stdout. Get more information on what Hadoop streaming is and how it works at http://hadoop.apache.org/docs/r1.2.1/streaming.html.

Getting ready

Refer to the *Executing our first sample MapReduce job using the mongo-hadoop connector* recipe to see how to set up Hadoop for development purposes and build the `mongo-hadoop` project using `gradle`. As far as Python libraries are concerned, we will install the required library from source. However, you can use `pip` to carry out the setup if you do not wish to build from source. We will also see how to set up `pymongo-hadoop` using `pip`.

Refer to the *Installing PyMongo* recipe in *Chapter 3, Programming Language Drivers*, to see how to install PyMongo and `pip`.

How it works...

1. We will first build `pymongo-hadoop` from source. With the project cloned to the local filesystem, execute the following commands from the root of the cloned project:

    ```
    $ cd streaming/language_support/python
    $ sudo python setup.py install
    ```

2. After you enter the password, setup will continue to install `pymongo-hadoop` on your machine.

3. That is all we need to do to build `pymongo-hadoop` from source. However, if you chose to not build from source, you could execute the following command from the operating system shell:

    ```
    $ sudo pip install pymongo_hadoop
    ```

4. After installing `pymongo-hadoop` in either way, we will now implement our `mapper` and `reducer` functions in Python.

5. The mapper function is as follows:

   ```
   #!/usr/bin/env python

   import sys
   from pymongo_hadoop import BSONMapper
   def mapper(documents):
     print >> sys.stderr, 'Starting mapper'
     for doc in documents:
       yield {'_id' : doc['state'], 'count' : 1}
     print >> sys.stderr, 'Mapper completed'
   ```

   BSONMapper(mapper)

6. The reducer function is as follows:

   ```
   #!/usr/bin/env python

   import sys
   from pymongo_hadoop import BSONReducer
   def reducer(key, documents):
     print >> sys.stderr, 'Invoked reducer for key "', key, '"'
     count = 0
     for doc in documents:
       count += 1
     return {'_id' : key, 'count' : count}
   ```

   BSONReducer(reducer)

7. The $HADOOP_HOME and $HADOOP_CONNECTOR_HOME environment variables should point to the base directory of Hadoop and the base directory of the mongo-hadoop connector project, respectively. Now, we will invoke the MapReduce function using the following command from the operating system shell. The code available on the book's website has the mapper and reducer Python script and a shell script that will be used to invoke mapper and reducer:

   ```
   $HADOOP_HOME/bin/hadoop jar \
   $HADOOP_HOME/share/hadoop/tools/lib/hadoop-streaming* \
   -libjars $HADOOP_CONNECTOR_HOME/streaming/build/libs/mongo-hadoop-streaming-1.2.1-SNAPSHOT-hadoop_2.4.jar \
   -input /tmp/in \
   -output /tmp/out \
   -inputformat com.mongodb.hadoop.mapred.MongoInputFormat \
   -outputformat com.mongodb.hadoop.mapred.MongoOutputFormat \
   -io mongodb \
   -jobconf mongo.input.uri=mongodb://127.0.0.1:27017/test.postalCodes \
   ```

Integration with Hadoop

```
-jobconf mongo.output.uri=mongodb://127.0.0.1:27017/test.
pyMRStreamTest \
-jobconf stream.io.identifier.resolver.class=com.mongodb.hadoop.
streaming.io.MongoIdentifierResolver \
-mapper mapper.py \
-reducer reducer.py
```

8. The `mapper.py` and `reducer.py` files are present in the current directory when executing this command.

9. On executing the command, which should take some time for successful execution of the MapReduce job, open the Mongo shell by typing the following command in the operating system command prompt and execute the following query from the shell:

    ```
    $ mongo
    > db.pyMRStreamTest.find().sort({count:-1}).limit(5)
    ```

10. Compare the output with the ones we got earlier when we executed the MapReduce jobs using Mongo's MapReduce framework in *Chapter 3, Programming Language Drivers*.

How to do it...

Let's look at steps 5 and 6 where we wrote the `mapper` and `reducer` functions. We defined a `map` function that accepts a list of all the documents. We iterated through these yield documents where the `_id` field is the name of the key, and the `value` field count has the value 1. The number of documents yielded will be the same as the total number of input documents.

Finally, we instantiated `BSONMapper`, which accepts the `mapper` function as the parameter. The function returned a generator object, which is then used by this `BSONMapper` class to feed the value to the MapReduce framework. All we need to remember is that that the `mapper` function needs to return a generator (which is returned as we call `yield` in the loop) and then instantiate the `BSONMapper` class, which is provided to us by the `pymongo_hadoop` module. Those intrigued enough might choose to look at the source code under the project cloned on your local filesystem in the `streaming/language_support/python/pymongo_hadoop/mapper.py` file and see what it does. It is a small piece of code that is simple to understand.

For the `reducer` function, we got the key and a list of documents for this key as the value. The key is the same as the value of the `_id` field emitted from the document in the `map` function. We simply returned a new document here with `_id` as the name of the state and `count` as the number of documents for this state. Remember that, here, we return a document and have not emitted one as we did in the `map` function. Again, finally, we instantiated `BSONReducer` and passed it to the `reducer` function. The source code under the project cloned on our local filesystem is in the `streaming/language_support/python/pymongo_hadoop/reducer.py` file, which has the implementation of the `BSONReducer` class file.

We finally invoked the command from the shell to initiate the MapReduce job that uses streaming. A few things to note here are that we need two JAR files: one in the `share/hadoop/tools/lib` directory of the Hadoop distribution and one in the `mongo-hadoop` connector, which is present in the `streaming/build/libs/` directory. The input and output formats are `com.mongodb.hadoop.mapred.MongoInputFormat` and `com.mongodb.hadoop.mapred.MongoOutputFormat`, respectively.

As we saw earlier, sysout and sysin form the backbone of streaming. So, basically, we need to encode our BSON objects to write to sysout; then, we should be able to read sysin to convert the content to BSON objects again. For this purpose, the `mongo-hadoop` connector provided us with two framework classes, `com.mongodb.hadoop.streaming.io.MongoInputWriter` and `com.mongodb.hadoop.streaming.io.MongoOutputReader`, to encode and decode from and to BSON objects, respectively. These classes extend from `org.apache.hadoop.streaming.io.InputWriter` and `org.apache.hadoop.streaming.io.OutputReader`.

The value of the `stream.io.identifier.resolver.class` property is given as `com.mongodb.hadoop.streaming.io.MongoIdentifierResolver`. This class extends from `org.apache.hadoop.streaming.io.IdentifierResolver` and gives us a chance to register our implementations of `org.apache.hadoop.streaming.io.InputWriter` and `org.apache.hadoop.streaming.io.OutputReader` with the framework. We also registered the output key and output value class using our custom `IdentifierResolver`. Just remember to use this resolver always if you are using streaming using the `mongo-hadoop` connector.

Finally, we gave the `mapper` and the `reducer` Python functions, which we discussed earlier. An important thing to remember is, do not print out logs to sysout from the `mapper` and `reducer` functions. The sysout and sysin are the means of communication, and writing logs to them can yield undesirable behavior. As we saw in the example, write to **standard error** (**stderr**) or, alternatively, write to a logfile.

When using a multiline command in Unix, you can continue the command on the next line using \. However, remember that there should be no spaces after \.

Integration with Hadoop

Running a MapReduce job on Amazon EMR

This recipe involves running the MapReduce job on the cloud using AWS. You will need an AWS account in order to proceed. Register to AWS at http://aws.amazon.com/. We will see how to run a MapReduce job on the cloud using Amazon **Elastic MapReduce** (**EMR**). Amazon EMR is a managed MapReduce service provided by Amazon on the cloud. For more details, refer to https://aws.amazon.com/elasticmapreduce/. Amazon EMR requires the data, binaries/jars, and so on to be present in the S3 bucket that it processes. It then writes the results back to the S3 bucket. Amazon **Simple Storage Service** (**S3**) is another service by AWS for data storage on the cloud. For more details on Amazon S3, refer to http://aws.amazon.com/s3/. Though we will use the mongo-hadoop connector, an interesting fact is that we won't require a MongoDB instance to be up and running. We will use the MongoDB data dump stored in an S3 bucket and use it for our data analysis. The MapReduce program will run on the input BSON dump and generate the resulting BSON dump in the output bucket. The logs of the MapReduce program will be written to another bucket dedicated to logs. The following diagram gives us an idea of how our setup will look like at a high level:

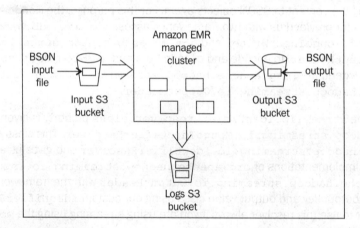

Getting ready

We will use the same Java sample for this recipe as the one we used in the *Writing our first Hadoop MapReduce job* recipe. To know more about the mapper and reducer class implementation, refer to the *How it works...* section of the *Writing our first Hadoop MapReduce job* recipe. We have a mongo-hadoop-emr-test project available with the code that can be downloaded from the book's website; this code is used to create a MapReduce job on the cloud using AWS EMR APIs.

Chapter 8

To simplify things, we will upload just one JAR file to the S3 bucket to execute the MapReduce job. This JAR file will be assembled using a BAT file for Windows and a shell script on Unix-based operating systems. The `mongo-hadoop-emr-test` Java project has a `mongo-hadoop-emr-binaries` subdirectory that contains the necessary binaries along with the scripts to assemble them into one JAR file. The assembled JAR file named `mongo-hadoop-emr-assembly.jar` is also provided in the subdirectory. Running the `.bat` or `.sh` file will delete this JAR file and regenerate the assembled JAR file; it is not mandatory to do this. The assembled JAR file that is already provided is good enough and will work just fine. The Java project contains a data subdirectory with a `postalCodes.bson` file in it. This is the BSON dump generated out of the database that contains the `postalCodes` collection. The `mongodump` utility provided with the Mongo distribution is used to extract this dump.

How to do it...

1. The first step of this exercise is to create a bucket on S3. You might choose to use an existing bucket. However, for this recipe, I am creating a bucket named `com.packtpub.mongo.cookbook.emr-in`. Remember that the name of the bucket has to be unique across all the S3 buckets; otherwise, you will not be able to create a bucket with this very name. You will have to create one with a different name and use it in place of `com.packtpub.mongo.cookbook.emr-in` used in this recipe.

   Do not create bucket names with an underscore (_); use a hyphen (-) instead. Bucket creation with an underscore will not fail, but the MapReduce job will fail later as it doesn't accept underscores in the bucket names.

2. We will upload the assembled JAR files and a `.bson` file for the data to the newly created (or existing) S3 bucket. To upload the files, we will use the AWS web console. Click on the **Upload** button and select the assembled JAR file and the `postalCodes.bson` file to be uploaded on the S3 bucket. After upload, the contents of the bucket will look like the following screenshot:

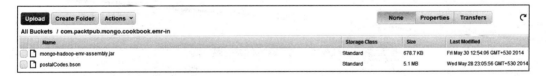

Integration with Hadoop

The following steps are to initiate the EMR job from the AWS console without writing a single line of code. We will also see how to initiate the same using the AWS Java SDK. Follow steps 4 to 9 if you are looking to initiate the EMR job from the AWS console. Follow steps 10 and 11 to start the EMR job using the Java SDK.

1. We will first initiate a MapReduce job from the AWS console. Visit `https://console.aws.amazon.com/elasticmapreduce/` and click on the **Create Cluster** button. In the **Cluster Configuration** screen, enter the details shown in the following screenshot, except for the logging bucket. You will need to select the bucket to which the logs need to be written. You might also click on the folder icon next to the textbox for the bucket name and select the bucket present for your account to be used as the logging bucket, as shown in the following screenshot:

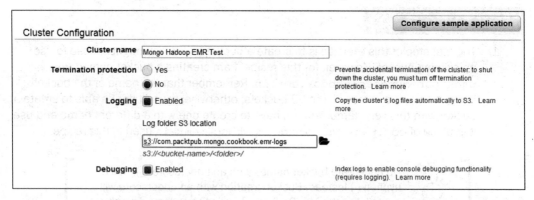

2. The **Termination protection** option is set to **No**, as this is a test instance. In the case of any error, we would prefer the instances to terminate to avoid keeping them running and incur charges.

3. In the **Software Configuration** section, select the Hadoop version as **2.4.0** and AMI version as **3.1.0**. Remove the additional applications by clicking on the cross next to their names, as shown in the following screenshot:

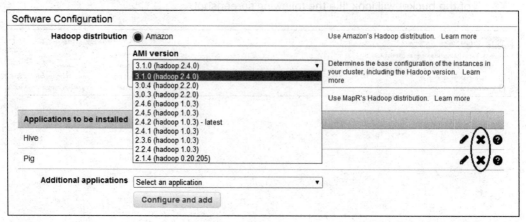

310

Chapter 8

4. In the **Hardware Configuration** section, select the EC2 instance type as **m1.medium**. This is the minimum we need to select for Hadoop Version 2.4.0. The number of instances for the slave and task instances is zero. The following screenshot shows the configuration selected:

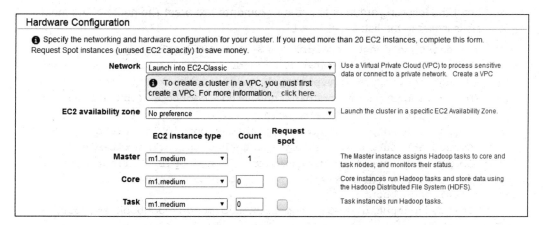

5. In the **Security and Access** section, leave all the default values. We also have no need for a Bootstrap Action, so leave this as is too.
6. The next step is to set up steps for the MapReduc job. In the **Add step** drop-down menu, select the **Custom JAR** option and select the **Auto-terminate** option to **Yes**, as shown in the following screenshot:

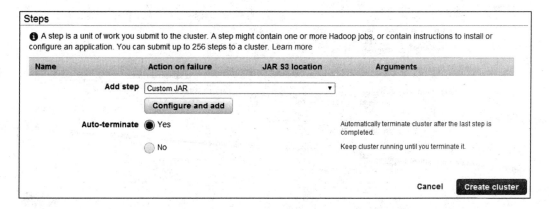

7. Now, click on the **Configure and add** button and enter the details.

Integration with Hadoop

8. The value of the **JAR S3 Location** field is given as **s3://com.packtpub.mongo.cookbook.emr-in/mongo-hadoop-emr-assembly.jar**. This is the location in my input bucket; you need to change the input bucket as per your own input bucket. The name of the JAR file will be same.

9. Enter the following arguments in the **Arguments** text area. The name of the main class is first in the list:

   ```
   com.packtpub.mongo.cookbook.TopStateMapReduceEntrypoint
   -Dmongo.job.input.format=com.mongodb.hadoop.BSONFileInputFormat
   -Dmongo.job.mapper=com.packtpub.mongo.cookbook.TopStatesMapper
   -Dmongo.job.reducer=com.packtpub.mongo.cookbook.TopStateReducer
   -Dmongo.job.output=org.apache.hadoop.io.Text
   -Dmongo.job.output.value=org.apache.hadoop.io.IntWritable
   -Dmongo.job.output.value=org.apache.hadoop.io.IntWritable
   -Dmongo.job.output.format=com.mongodb.hadoop.BSONFileOutputFormat
   -Dmapred.input.dir=s3://com.packtpub.mongo.cookbook.emr-in/postalCodes.bson
   -Dmapred.output.dir=s3://com.packtpub.mongo.cookbook.emr-out/
   ```

10. Again, the value of the final two arguments contains the input and output buckets used for my MapReduce sample. This value will change according to your own input and output buckets. The value of **Action on failure** will be **Terminate cluster**. The following screenshot shows the values filled. Click on **Save** after all the preceding details are entered in:

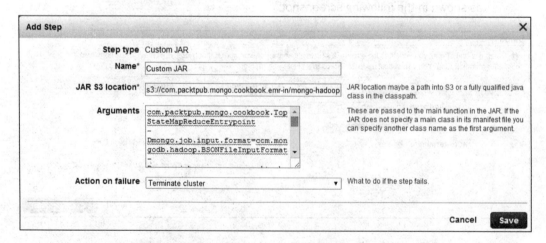

11. Now, click on the **Create Cluster** button. This will take some time to provision and start the cluster.

Chapter 8

12. In the following few steps, we will create a MapReduce job on EMR using the AWS Java API. Import the `EMRTest` project provided with the code samples into your favorite IDE. Once imported, open the `com.packtpub.mongo.cookbook.AWSElasticMapReduceEntrypoint` class.

13. There are five constants that need to be changed in the class. They are input; output; log bucket, which you will use for your example; the EC2 key name; the AWS access; and the secret key. The access key and secret key act as the user name and password, respectively, when you use AWS SDK. Change these values accordingly and run the program; on successful execution, it should give you a job ID for the newly initiated job.

14. Irrespective of how you initiated the EMR job, visit the EMR console at `https://console.aws.amazon.com/elasticmapreduce/` to see the status of your submitted ID. The job ID you see in the second column of your initiated jobs will be the same as the job ID printed to the console when you executed the Java program (if you initiated the job using the Java program). Click on the name of the job initiated; this should navigate you to the job-details page. The hardware provisioning will take some time and then, finally, your MapReduce step will run. Once the job is complete, the status of the job will look like the following screenshot:

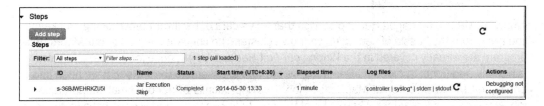

15. When the **Steps** section is expanded, it will look like the following screenshot:

ID	Name	Status	Start time (UTC+5:30)	Elapsed time	Log files	Actions
s-36BJWEHRKZU5I	Jar Execution Step	Completed	2014-05-30 13:33	1 minute	controller | syslog | stderr | stdout	Debugging not configured

16. Click on the **stderr** link below the **Log files** section to view all the logs' output for the MapReduce job.

17. Now that the MapReduce job is complete, our next step is to see the results of it. Visit the S3 console at `https://console.aws.amazon.com/s3`, and visit the out bucket set. In my case, the following is the content of the out bucket:

Name	Storage Class	Size	Last Modified
_SUCCESS	Standard	0 bytes	Fri May 30 13:34:28 GMT+530 2014
part-r-00000.bson	Standard	1.2 KB	Fri May 30 13:34:27 GMT+530 2014

313

Integration with Hadoop

The `part-r-0000.bson` file interests us. This file contains the results of our MapReduce job.

18. Download the file to your local filesystem and import it into a running Mongo instance locally using the `mongorestore` utility as follows. Note that the restore utility for the following command expects a `mongod` instance to be up and running and listening to port `27017` and the `part-r-0000.bson` file in the current directory:

    ```
    $ mongorestore part-r-00000.bson -d test -c mongoEMRResults
    ```

19. Now, connect to the `mongod` instance using the Mongo shell and execute the following query:

    ```
    > db.mongoEMRResults.find().sort({count:-1}).limit(5)
    { "_id" : "Maharashtra", "count" : 6446 }
    { "_id" : "Kerala", "count" : 4684 }
    { "_id" : "Tamil Nadu", "count" : 3784 }
    { "_id" : "Andhra Pradesh", "count" : 3550 }
    { "_id" : "Karnataka", "count" : 3204 }
    ```

20. The preceding command shows the top five results. If we compare the results we got in *Chapter 3, Programming Language Drivers*, for using Mongo's MapReduce framework and the *Writing our first Hadoop MapReduce job* recipe in this chapter, we will see that the results are identical.

How it works...

Amazon EMR is a managed Hadoop service that takes care of hardware provisioning and keeps you away from the hassle of setting up your own cluster. The concepts related to our MapReduce program are already covered in the *Writing our first Hadoop MapReduce job* recipe, and there is nothing additional to mention. One thing we did was to assemble the JARs that we need into one big fat JAR to execute our MapReduce job. This approach is OK for our small MapReduce job. In the case of larger jobs where a lot of third-party JARs are needed, we will have to go for an approach where we will add the JARs to the `lib` directory of the Hadoop installation and execute it in the same way we did in our MapReduce job that we executed locally. Another thing that we did differently from what we did in our local setup was to not use a `mongod` instance to source the data and write the data; instead, we used BSON dump files from the Mongo database as an input and write the output to BSON files. The output dump will then be imported to a Mongo database locally, and the results will be analyzed. It is pretty common to have data dumps uploaded to S3 buckets; thus, running analytics jobs on this data uploaded to S3 on the cloud using cloud infrastructure is a good option. The data accessed from the buckets by the EMR cluster need not have public access, as the EMR job runs using our account's credentials; we are good to access our own buckets to read and write data/logs.

See also

- The developer's guide for EMR at `http://docs.aws.amazon.com/ElasticMapReduce/latest/DeveloperGuide/`.
- `https://github.com/mongodb/mongo-hadoop/tree/master/examples/elastic-mapreduce` to see the sample MapReduce job on the `enron` dataset given as part of the `mongo-hadoop` connector's examples. You might choose to implement this example on Amazon EMR as per the given instructions.

9
Open Source and Proprietary Tools

In this chapter, we will cover some open source and proprietary tools. We will cover the following recipes in this chapter:

- Developing using spring-data-mongodb
- Accessing MongoDB using Java Persistence API
- Accessing MongoDB over REST
- Installing the GUI-based client, MongoVUE, for MongoDB

Introduction

There is a vast array of tools/frameworks available to ease the development/administration process for software that uses MongoDB. We will look at some of these available frameworks and tools. For developer productivity (Java developers in this case), we will look at `spring-data-mongodb`, which is part of the popular Spring data suite.

Java Persistence API (**JPA**) is an **object relational mapping** (**ORM**) specification that is widely used, particularly with relational databases (this was the objective of ORM frameworks). However, there are few implementations that let us use it with NoSQL stores, MongoDB in this case. We will look at one provider that provides this implementation and put it to the test with a simple use case.

Open Source and Proprietary Tools

We will use `spring-data-rest` to expose CRUD repositories for MongoDB over a REST interface for clients to invoke various operations supported by the underlying `spring-data-mongo` repository.

Querying the database from the shell is OK, but it will be nice to have a good GUI to enable us to do all administrative/development-related tasks from the GUI rather than executing the commands from the shell to perform these activities. We will look at one such tool, MongoVUE, in this chapter.

Developing using spring-data-mongodb

From the perspective of developers, when a program needs to interact with a MongoDB instance, they need to use the respective client APIs for their specific platforms. The trouble with doing this is that we need to write a lot of boilerplate code, and it is not necessarily object-oriented. For instance, we have a class called `Person` with various attributes such as `name`, `age`, and `address`. The corresponding JSON document too shares a similar structure to this `Person` class as follows:

```
{
    name:"…",
    age:..,
    address:{lineOne:"…", …}
}
```

However, to store this document, we need to convert the `Person` class to a `DBObject`, which is a map with key and value pairs. What is really desired is to let us persist this `Person` class itself as an object in the database, without having to convert it to `DBObject`.

Also, some of the operations such as searching by a particular field of a document, saving an entity, deleting an entity, and searching by ID are pretty common, and we tend to repeatedly write similar boilerplate code. In this recipe, we will see how `spring-data-mongodb` relieves us of these laborious and cumbersome tasks not only to reduce the development effort but also to dramatically reduce the possibility of introducing bugs in these commonly written functions.

Getting ready

The `SpringDataMongoTest` project present in the bundle in the chapter is a Maven project and is to be imported into any IDE of your choice. The required Maven artifacts will automatically be downloaded. A single MongoDB instance that listens to port `27017` is required to be up-and-running. Refer to the *Single node installation of MongoDB* recipe in *Chapter 1, Installing and Starting the MongoDB Server*, to know how to start a standalone instance.

For the aggregation example, we will use the postal code data. Refer to the *Creating test data* recipe in *Chapter 2, Command-line Operations and Indexes*, for the creation of test data.

How to do it...

1. We will explore the repository feature of spring-data-mongodb first. Open the test case class named `com.packtpub.mongo.cookbook.MongoCrudRepositoryTest` from your IDE and execute it. If all goes well and the MongoDB server instance is reachable, the test case will execute successfully.

2. Another test case named `com.packtpub.mongo.cookbook.MongoCrudRepositoryTest2` is used to explore more features of the repository support provided by `spring-data-mongodb`. This test case too should get executed successfully.

3. We will see how `MongoTemplate` of spring-data-mongodb can be used to perform CRUD operations and other common operations on MongoDB. Open the `com.packtpub.mongo.cookbook.MongoTemplateTest` class and execute it.

4. Alternatively, if an IDE is not used, all the tests can be executed using Maven from the command prompt as follows, with the current directory being the root of the `SpringDataMongoTest` project:

   `$ mvn clean test`

How it works...

We will first look at what we did in `com.packtpub.mongo.cookbook.MongoCrudRepositoryTest`, where we saw the repository support provided by `spring-data-mongodb`. Just in case you didn't notice, we haven't written a single line of code for the repository. The magic of implementing the required code for us is done by the Spring data project. Let's start by looking at the relevant portions of the XML `config` file:

```
<mongo:repositories base-package="com.packtpub.mongo.cookbook" />
<mongo:mongo id="mongo" host="localhost" port="27017"/>
<mongo:db-factory id="factory" dbname="test" mongo-ref="mongo"/>
<mongo:template id="mongoTemplate" db-factory-ref="factory"/>
```

We will first look at the last three lines: `spring-data-mongodb` namespace declarations to instantiate `com.mongodb.Mongo`, instantiating a factory for the `com.mongodb.DB` instances from the client, and a `template` instance, used to perform various operations on MongoDB, respectively. We will see `org.springframework.data.mongodb.core.MongoTemplate` in more detail later.

The first line is a namespace declaration for the base package of all the CRUD repositories we have. In this package, we have an interface with the following body:

```
public interface PersonRepository extends PagingAndSortingRepository<Person, Integer>{

    /**
     *
     * @param lastName
     * @return
     */
    Person findByLastName(String lastName);
}
```

The `PagingAndSortingRepository` interface is from the `org.springframework.data.repository` package of the Spring data core project and extends from `CrudRepository` of the same project. These interfaces give us some most common methods such as searching by the ID/primary key, deleting an entity, inserting an entity, and updating an entity. The repository needs an object that it maps to the underlying data store. The Spring data project supports a large number of data stores that are not just limited to SQL (using **Java Database Connectivity** (**JDBC**) and JPA) or MongoDB but also to other NoSQL stores such as Redis and Hadoop and search engines such as Solr and Elasticsearch. In the case of `spring-data-mongodb`, the object is mapped to a document in the collection.

The `PagingAndSortingRepository<Person, Integer>` signature indicates that the first one is the entity that the CRUD repository is built for, and the second one is the type of the primary key/ID field.

We have added just one method named `findByLastName`; this accepts one string value for the last name as a parameter. This is an interesting operation; it is specific to our repository and not even implemented by us, but it will still work just as expected. The `Person` class is a POJO where we have annotated the ID field with the `org.springframework.data.annotation.Id` annotation. Nothing else is really special about this class; it just has some plain getters and setters.

With all these small details, let's join these dots together by answering some questions you'll have in mind. First, we will see which server, database, and collection our data goes to. If we look at the `mongo:mongo` XML definition for the `config` file, we will see that we instantiated the `com.mongodb.Mongo` class by connecting to a localhost and port `27017`. The `mongo:db-factory` declaration is used to denote that the database to be used is `test`. One final question is, which collection? The simple name of our class is `Person`. The name of the collection is the simple name with the first character lowercased; thus, `Person` will become `person`, and something like `BillingAddress` will become the `billingAddress` collection. These are the default values. However, if you need to override this value, you can annotate your class with the `org.springframework.data.mongodb.core.mapping.Document` annotation and use its attribute collection to give any name of your choice, as we will see in an example later.

To view the document in the collection, execute just one test case method called `saveAndQueryPerson` from the `com.packtpub.mongo.cookbook.MongoCrudRepositoryTest` class. Now, connect to the MongoDB instance from the Mongo shell and execute the following query:

```
> use test
> db.person.findOne({_id:1})
{
  "_id" : 1,
  "_class" : "com.packtpub.mongo.cookbook.domain.Person",
  "firstName" : "Steve",
  "lastName" : "Johnson",
  "age" : 20,
  "gender" : "Male"
  ...
}
```

As we can see in the preceding result, the contents of the document are similar to the object we persisted using the CRUD repository. The names of the field in the document are the same as the names of the respective attributes in the Java object, with two exceptions. The field annotated with `@Id` is now `_id`, irrespective of the name of the field in the Java class, and an additional `_class` attribute is added to the document whose value is the fully qualified name of the Java class itself. This is not of any use for the application but is used by spring-data-mongodb as metadata.

Now, it makes more sense and gives us an idea about what spring-data-mongodb must be doing for all the basic CRUD methods. All the operations we perform will use the `MongoTemplate` class (`MongoOperations` to be precise, which is an interface that `MongoTemplate` implements) from the spring-data-mongodb project. To find it by using the primary key, it will invoke a `find` query by the `_id` field on the collection derived using the `Person` entity class. The save method simply calls the `save` method on `MongoOperations`; this in turn calls the `save` method on the `com.mongodb.DBCollection` class.

We still haven't answered how the `findByLastName` method worked. How does Spring know what query to invoke to return the data? These are the special types of methods that begin with `find`, `findBy`, `get`, or `getBy`. There are some rules that one needs to follow while naming a method, and the proxy object on the repository interface is able to correctly convert this method into an appropriate query on the collection. For instance, the `findByLastName` method in the repository of the `Person` class will execute a query on the `lastName` field in the document of the `Person` class. Hence, the `findByLastName(String lastName)` method will fire the following query in the database:

`db.person.find({'lastName': lastName })`

Open Source and Proprietary Tools

Based on the return type of the method defined, it will return either a list or the first result in the returned result from the database. We have used `findBy` in our queries. However, for anything that begins with `find`, having any text in between and having text that ends in `By` works. For instance, `findPersonBy` is the same as `findBy`.

For more information on these `findBy` methods, we have another test class, `MongoCrudRepositoryTest2`. Open this class in your IDE where it can be read along with this text. We have already executed this test case; now, let's see these `findBy` methods used and their behavior. This class has seven `findBy` methods in it, with one of the methods being a variant of another method in the same interface. To get a clear idea of the queries, we will first look at one of the documents in the `personTwo` collection in the `test` database. Execute the following commands on the Mongo shell connected to the MongoDB server that runs on a localhost:

```
> use test
> db.personTwo.findOne({firstName:'Amit'})
{
  "_id" : 2,
  "_class" : "com.packtpub.mongo.cookbook.domain.Person2",
  "firstName" : "Amit",
  "lastName" : "Sharma",
  "age" : 25,
  "gender" : "Male",
  "residentialAddress" : {
  "addressLineOne" : "20, Central street",
  "city" : "Mumbai",
  "state" : "Maharashtra",
  "country" : "India",
  "zip" : "400101"
  }
}
```

Also, note that the repository uses the `Person2` class. However, the name of the collection used is `personTwo`. This was possible because we used the `@Document(collection="personTwo")` annotation on top of the `Person2` class.

Getting back to the seven methods in the `com.packtpub.mongo.cookbook.PersonRepositoryTwo` repository class, let's look at them one by one:

Method	Description
`findByAgeGreaterThanEqual`	This method will fire the `{'age':{'$gte':<age>}}` query on the `personTwo` collection.
	The secret lies in the name of the method. If we break it up, what we have after `findBy` tells us what we want. The `age` property (with first character lowercased) is the field that would be queried on the document with the `$gte` operator because we have `GreaterThanEqual` in the name of the method. The value that would be used for the comparison would be the value of the parameter passed. The result is a collection of `Person2` entities, as we will have multiple matches.
`findByAgeBetween`	This method will query on `age` but will use a combination of `$gt` and `$lt` to find the matching result. The query in this case will be `{'age' : {'$gt' : from, '$lt' : to}}`. It is important to note that both the `from` and `to` values are exclusive in the range. There are two methods in the test case: `findByAgeBetween` and `findByAgeBetween2`. These methods demonstrate the behavior of the between query for different input values.
`findByAgeGreaterThan`	This method is a special method that also sorts the result because there are two parameters to the method: the value against which the `age` parameter will be compared is the first parameter, and the second parameter is the field of type `org.springframework.data.domain.Sort`. For more details, refer to the Javadoc for `spring-data-mongodb`.
`findPeopleByLastNameLike`	This method is used to find results by the last name that matches a pattern. Regular expressions are used for the matching purpose. For instance, in this case, the query fired will be `{'lastName' : <lastName as regex>}`. This method's name begins with `findPeopleBy` instead of `findBy`, which works in the same way as `findBy`. Thus, when we say `findBy` in all the descriptions, we actually mean find...By.
	The value provided as the parameter will be used to match the last name.

Method	Description
`findByResidential AddressCountry`	This is an interesting method to look at. Here, we are looking to search by the country of the residential address. This is, in fact, a field in the `Address` class in the `residentialAddress` field of the person. Take a look at the document from the `personTwo` collection mentioned earlier for how the query will be used. When the Spring data finds the name as `ResidentialAddressCountry`, it will try to find various combinations using this string. For instance, it can look at `residentialAddressCountry` in `Person` or in `residential.addressCountry`, `residentialAddress.country`, or `residential.address.country`. If there are no conflicting values, as in our case, `residentialAddress.country` is a success path in the `Person2` object tree, and thus, this will be used in the query. However, if there are conflicts, then underscores can be used to clearly specify what we are looking at. In this case, the method can be renamed to `findByResidentialAddress_country` to clearly specify what we expect as the result. The `findByCountry2` test case method demonstrates this.
`findByFirstName AndCountry`	This is an interesting method as well. We are not always able to use the method names to implement what we actually want to, as the name of the method required for Spring to automatically implement the query might be bit awkward to use as is. For instance, `findByCountryOfResidence` sounds better than `findByResidentialAddressCountry`. However, we are stuck with the latter, as this is how `spring-data-mongodb` will construct the query. Using `findByCountryOfResidence` gives no details on how to construct the query to Spring data. However, there is a solution to this. You might choose to use the `@Query` annotation and specify the query to be executed when the method is invoked. The following is the annotation we used in our case: `@Query("{'firstName':?0, 'residentialAddress.country': ?1}")` We write the value as a query that will get executed, and we bind the parameters of the functions to the query as numbered parameters that start from 0. Thus, the first parameter of the method will be bound to ?0, the second to ?1, and so on.

We saw how the `findBy` or `getBy` method is automatically translated to queries for MongoDB. Similarly, we have some well-known prefixes for the methods. The `countBy` prefix returns the long number for the count for a given condition, which is derived from the rest of the method names that are similar to `findBy`. We can have `deleteBy` or `removeBy` to delete the documents by the derived condition. Also, one thing to note about the `com.packtpub.mongo.cookbook.domain.Person2` class is that it does not have a no-argument constructor or setter to set the values. Spring will, instead, use reflection to instantiate this object.

A lot of `findBy` methods are supported by `spring-data-mongodb`, and all are not covered here. For more details, refer to the `spring-data-mongodb` reference manual. A lot of XML-based or Java-based configuration options are available too and can be found in the reference manual. The links are given in the *See also* section later in this recipe.

We are not done yet, though; we have another test case, `com.packtpub.mongo.cookbook.MongoTemplateTest`. This uses `org.springframework.data.mongodb.core.MongoTemplate` to perform various operations. We can open the test case class, and we will see what operations are performed and which methods of the `MongoTemplate` class are invoked.

Let's look at some of the important and frequently used methods of the `MongoTemplate` class:

Method	Description
save	This method is used to save (insert if new; otherwise, update) an entity in MongoDB. The method takes one parameter, the entity, and finds the target collection based on its name or the `@Document` annotation present in it. There is an overloaded version of the `save` method that also accepts the second parameter, which is the name of the collection to which the data entity passed needs to be persisted.
remove	This method will be used to remove documents from the collection. It has got some overloaded versions in this class. All of them accept either an entity to be deleted or the `org.springframework.data.mongodb.core.query.Query` instance, which is used to determine the document(s) to be deleted. The second parameter is then the name of the collection from which the document is to be deleted. When an entity is provided, the name of the collection can be derived. With a `Query` instance provided, we have to give either the name of the collection or the entity class name, which in turn will be used to derive the name of the collection.

Method	Description
`updateMulti`	This is the function invoked to update multiple documents with one update call. The first parameter is the query that will be used to match the documents. The second parameter is an instance of `org.springframework.data.mongodb.core.query.Update`. This is the update that will be executed on the documents selected using the first query object. The next parameter is the entity class or the collection name to execute the update on. For more details on the method and its various overloaded versions, refer to the Javadoc.
`updateFirst`	This is the opposite of the `updateMulti` method. This operation will update just the first matching document. We have not covered this method in our unit test case.
`insert`	We mentioned that the `save` method can perform insertions and updates. The `insert` method in the template calls the `insert` method of the underlying Mongo client. If one entity document is to be inserted, there is no difference in calling the `insert` or `save` method. However, as we saw in the test case in the `insertMultiple` method, we created a list of three `Person` instances and passed them to the `insert` method. All the three documents for the three `Person` instances will go to the server as part of one call. The behavior of an insert when it fails is determined by the `continue on error` parameter of the write concern. It will determine whether the bulk insert fails on the first failure or continues even after errors that report only the last error. The page at `http://docs.mongodb.org/manual/core/bulk-inserts/` gives more details on bulk inserts and various write concern parameters that can alter the behavior.
`findAndRemove/ findAllAndRemove`	Both these operations are used to find and then remove the document(s). The `findAndRemove` method finds one document and then returns the deleted document. This operation is atomic. However, the `findAllAndRemove` method finds all the documents and removes them before returning the list of all the entities of all the documents deleted.

Method	Description
findAndModify	This method is functionally similar to findAndModify that we have with the Mongo client library. It will atomically find and modify the document. If the query matches more than one document, only the first match will be updated. The first two parameters of this method are the query and the update to execute. The next parameter is either the entity class or the collection name to execute the operation on. Also, there is a special class called org.springframework.data.mongodb.core.FindAndModifyOptions that makes sense only for the findAndModify operation. This instance tells us whether we are looking for the new instance or the old instance after the operation is performed, and whether the upsert operation is to be performed and it is relevant only if the document with the matching query doesn't exist. There is an additional Boolean flag to tell the client whether this is a find and remove operation. In fact, the findAndRemove operation we saw earlier is just a convenience function that delegates to findAndModify with this remove flag set.

In the preceding table, we mentioned the Query and Update classes when talking about update. These are special convenience classes in spring-data-mongodb; they let us build MongoDB queries using a syntax that is easy to understand and improves readability. For instance, the query to check whether lastName is Johnson in the Mongo query language is {'lastName':'Johnson'}. The same query can be constructed in spring-data-mongodb as follows:

new Query(Criteria.where("lastName").is("Johnson"))

This syntax looks neat as compared to giving the query in JSON. Let's take another example where we want to find all females under 30 years of age in our database. The query will now be built as follows:

new Query(Criteria.where("age").lt(30).and("gender").is("Female"))

Similarly, for update, we want to set a youngCustomer Boolean flag to be true for some of the customers, based on some conditions. To set this flag in the document, the MongoDB format will be as follows:

{'$set' : {'youngCustomer' : true}}

In spring-data-mongodb, the same will be achieved as follows:

new Update().set("youngCustomer", true)

Refer to the Javadoc for all the possible methods that are available to build the query and updates in spring-data-mongodb that are to be used with MongoTemplate.

Open Source and Proprietary Tools

The methods mentioned earlier are by no means the only ones available in the `MongoTemplate` class. There are a lot of other methods for geospatial indexes, convenience methods to get the count of documents in the collection, aggregation, and MapReduce support, and so on. Refer to the Javadoc of `MongoTemplate` for more details and methods.

Speaking of aggregation, we also have a test case method called `aggregationTest` to perform the `aggregation` operation on the collection. We have a `postalCodes` collection in MongoDB; this collection contains the postal code details of various cities. An example document in the collection is as follows:

```
{
  "_id" : ObjectId("539743b26412fd18f3510f1b"),
  "postOfficeName" : "A S D Mello Road Fuller Marg",
  "pincode" : 400001,
  "districtsName" : "Mumbai",
  "city" : "Mumbai",
  "state" : "Maharashtra"
}
```

Our `aggregation` operation intends to find the top five states by the number of documents in the collection. In Mongo, the aggregation pipeline will look as follows:

```
[
{'$project':{'state':1, '_id':0}},
{'$group':{'_id':'$state', 'count':{'$sum':1}}}
{'$sort':{'count':-1}},
{'$limit':5}
]
```

In `spring-data-mongodb`, we invoked the `aggregation` operation using the `MongoTemplate` class as follows:

```
Aggregation aggregation = newAggregation(

  project("state", "_id"),
  group("state").count().as("count"),
  sort(Direction.DESC, "count"),
  limit(5)
);


AggregationResults<DBObject> results = mongoTemplate.aggregate(
```

```
aggregation,
"postalCodes",
DBObject.class);
```

The key is in creating an instance of the `org.springframework.data.mongodb.core.aggregation.Aggregation` class. The `newAggregation` method is statically imported from the same class and accepts varargs for different instances of `org.springframework.data.mongodb.core.aggregation.AggregationOperation` that correspond to the one operation in the chain. The `Aggregation` class has various static methods to create the instances of `AggregationOperation`. We have used a few of them, such as `project`, `group`, `sort`, and `limit`. For more details and available methods, refer to the Javadoc. The `aggregate` method in `MongoTemplate` takes three arguments. The first one is the instance of the `Aggregation` class, the second one is the name of the collection, and the third one is the return type of the aggregation result. Refer to the `aggregation` operation test case for more details.

See also

- The Javadoc at `http://docs.spring.io/spring-data/mongodb/docs/current/api/` for more details and API documentation
- The reference manual for the `spring-data-mongodb` project at `http://docs.spring.io/spring-data/data-mongodb/docs/current/reference/`

Accessing MongoDB using Java Persistence API

In this recipe, we will use a JPA provider that allows us to use JPA entities to achieve object-to-document mapping with MongoDB.

Getting ready

Start the standalone server instance that listens to port `27017`. This is a Java project using JPA. Familiarity with JPA and its annotations is expected, though what we will be looking at is fairly basic. Refer to the *Connecting to a single node from a Java client* recipe in *Chapter 1, Installing and Starting the MongoDB Server*, to know how to set up Maven if you are not aware of it. Download the `DataNucleusMongoJPA` project from the code bundle provided with the book. Though we will execute the test cases from the command prompt, you may import the project in your favorite IDE to view the source code.

Open Source and Proprietary Tools

How to do it...

1. Go to the root directory of the `DataNucleusMongoJPA` project and execute the following command in the shell:

   `$ mvn clean test`

2. This will download the necessary artifacts needed to build and run the project; then, execute the test cases successfully.

3. Once the test cases get executed, open a Mongo shell and connect to the local instance.

4. Execute the following query in the shell:

   ```
   > use test
   > db.personJPA.find().pretty()
   ```

How it works...

First, let's look at a sample document that got created in the `personJPA` collection:

```
{
  "_id" : NumberLong(2),
  "residentialAddress" : {
  "residentialAddress_zipCode" : "400101",
  "residentialAddress_state" : "Maharashtra",
  "residentialAddress_country" : "India",
  "residentialAddress_city" : "Mumbai",
  "residentialAddress_addressLineOne" : "20, Central street"
},
  "lastName" : "Sharma",
  "gender" : "Male",
  "firstName" : "Amit",
  "age" : 25
}
```

The steps we executed are pretty simple; let's look at the classes used one by one. We will start with the `com.packtpub.mongo.cookbook.domain.Person` class. At the top of the class (after the package and imports), we have the following:

```
@Entity
@Table(name="personJPA")
public class Person {
```

This denotes that the `Person` class is an entity, and the collection to which it will persist is `personJPA`. Note that JPA was designed primarily as an ORM tool; thus, the terminologies used are more for a relational database. A table in RDBMS is synonymous with a collection in MongoDB. The rest of the class contains the attributes of the person and the columns annotated with `@Column` and `@Id` for the primary key. These are simple JPA annotations. What is interesting to look at is the `com.packtpub.mongo.cookbook.domain.ResidentialAddress` class, which is stored as the `residentialAddress` variable in the `Person` class. If we look at the person document we gave earlier, all the values given in the `@Column` annotation are the names of the keys for person. Also, notice how the enum too gets converted to the string value. However, the `residentialAddress` field is the name of the variable in the `Person` class against which the address instance is stored. If we look at the `ResidentialAddress` class, we will see the `@Embeddable` annotation on top, above the class name. This is again a JPA annotation that denotes that this instance is not an entity unto itself but is embedded in another entity or another embeddable class. Note the names of the fields in the document. They have the following format:

```
<name of the variable in person class>_<value of the variable name in
ResidentialAddress class>
```

There is one problem we notice here. The names of the fields are too long, thus consuming unnecessary space. The solution is to have a shorter value in the `@Column` annotation. For instance, have the following annotation:

```
@Column(name="ln") instead of @Column(name="lastName")
```

This will create the key with the name `ln` in the document. Unfortunately, this doesn't work with the embedded `ResidentialAddress` class; in this case you will have to deal with shorter variable names. Now that we have seen the entity classes, let's look at the `persistence.xml` file:

```xml
<persistence-unit name="DataNucleusMongo">
  <class>com.packtpub.mongo.cookbook.domain.Person</class>
  <properties>
     <property name="javax.persistence.jdbc.url"
value="mongodb:localhost:27017/test"/>
   </properties>
</persistence-unit>
```

We have just got the `persistence-unit` definition here, with the name as `DataNucleusMongo`. There is one class node, which is the entity that we will use. Note that the embedded address class is not mentioned here as it is not an independent entity. In the properties, we mentioned the URL of the data store to connect to. In this case, we connected to the instance on the localhost, which is port `27017`, and the `test` database.

Open Source and Proprietary Tools

Now, let's look at the class that queries and inserts the data. This is our test class, `com.packtpub.mongo.cookbook.DataNucleusJPATest`. We will create `javax.persistence.EntityManagerFactory` as `Persistence.createEntityManagerFactory("DataNucleusMongo")`. This is a thread-safe class, and its instance is shared across threads. Also, the string argument is the same as the name of the persistence unit we used in `persistence.xml`. All other invocations on `javax.persistence.EntityManager` to persist or query the collection require us to create an instance using `EntityManagerFactory`, use it, and then close it once the operation is complete. All the operations performed are as per the JPA specification. The test case class persists entities and also queries them.

Finally, we will look at `pom.xml`, particularly, the enhancer plugin we used; it is as follows:

```xml
<plugin>
  <groupId>org.datanucleus</groupId>
  <artifactId>datanucleus-maven-plugin</artifactId>
  <version>4.0.0-release</version>
  <configuration>
    <log4jConfiguration>${basedir}/src/main/resources/log4j.properties</log4jConfiguration>
    <verbose>true</verbose>
  </configuration>
  <executions>
    <execution>
      <phase>process-classes</phase>
      <goals>
        <goal>enhance</goal>
      </goals>
    </execution>
  </executions>
</plugin>
```

The entities we wrote need to be enhanced to be used as JPA entities that use a data nucleus. This preceding plugin will be attached to the `process-classes` phase and then call the plugin's enhance.

See also

- `http://www.datanucleus.org/products/datanucleus/jdo/enhancer.html` for possible options. There is also a plugin for Eclipse to allow entity classes to be enhanced/instrumented for a data nucleus.
- The JPA 2.1 specification at `https://www.jcp.org/aboutJava/communityprocess/final/jsr338/index.html`.

Accessing MongoDB over REST

In this recipe, we will see how to access MongoDB and perform CRUD operations using REST APIs. We will use `spring-data-rest` for REST access and `spring-data-mongodb` to perform the CRUD operations. Before you continue with this recipe, it is important to know how to implement CRUD repositories using `spring-data-mongodb`. Refer to the *Developing using spring-data-mongodb* recipe to know how to use this framework.

The question that one must have is, why a REST API is needed? There are scenarios where there is a database that is being shared by many applications, possibly written in different languages. Writing JPA DAO or using `spring-data-mongodb` is good enough for Java clients, but not for clients in other languages. Having APIs locally with the application doesn't even give us a centralized way to access the database. This is where REST APIs come into play. We can develop the server-side data access layer, which is the CRUD repository in Java (`spring-data-mongodb` to be precise), and then expose it over a REST interface for a client written in any language to invoke it. Now, we will invoke our API in a platform-independent way and this will also give us a single point of entry into our database.

Getting ready

Apart from the prerequisites of the *Developing using spring-data-mongodb* recipe, we have a few more for this recipe. The first thing is to download the `SpringDataRestTest` project from the book's website and import it into your IDE as a Maven project. Alternatively, if you do not wish to import into the IDE, you can run the server that services the requests from the command prompt, which we will see in the next section. There is no specific client application used to perform the CRUD operations over REST. I will demonstrate the concepts using the Chrome browser and use a special plugin of the browser called **Advanced REST Client** to send HTTP POST requests to the server. The tools can be found under the **Developer Tools** section on the Chrome web store.

How to do it...

1. If you have imported the project in your IDE as a Maven project, execute the `com.packtpub.mongo.cookbook.rest.RestServer` class, which is the bootstrap class, and locally start the server that will accept client connections.

2. If the project is to be executed from the command prompt as a Maven project, go to the root directory of the project and run the following command:

   `mvn spring-boot:run`

3. The following output will be seen in the command prompt if all goes well and the server is started:

   `[INFO] Attaching agents: []`

Open Source and Proprietary Tools

4. After starting the server in either way, enter `http://localhost:8080/people` in the browser's address bar, and we will see the following JSON response. The following response is seen because the underlying collection, `person`, is empty:

```
{
  "_links" : {
    "self" : {
      "href" : "http://localhost:8080/people{?page,size,sort}",
      "templated" : true
    },
    "search" : {
      "href" : "http://localhost:8080/people/search"
    }
  },
  "page" : {
    "size" : 20,
    "totalElements" : 0,
    "totalPages" : 0,
    "number" : 0
  }
}
```

5. We will now insert a new document in the `person` collection using the HTTP POST request to `http://localhost:8080/people`. We will send a POST request to the server using the **Advanced REST Client** chrome extension. The document posted is as follows:

`{"lastName":"Cruise", "firstName":"Tom", "age":52, "id":1}`

The request's content type is `application/json`

Chapter 9

6. The following screenshot shows the `POST` request sent to the server and the response from the server:

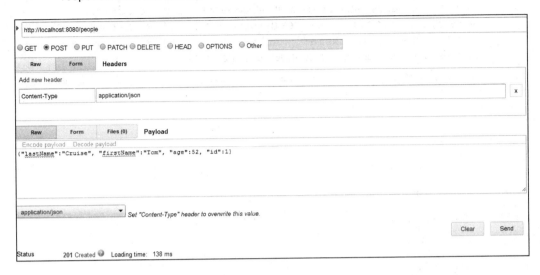

7. We will now query this document from the browser using the `_id` field, which is `1` in this case. Enter `http://localhost:8080/people/1` in the browser's address bar. You will see the document we inserted in step 5.

8. Now that we have one document in the collection (you might try to insert more documents for people with different names and, more importantly, a unique ID), we will query the document using the last name. However, first type in `http://localhost:8080/people/search` in the browser's address bar to view all the search options available. We will see one search method, `findByLastName`, that accepts a command-line parameter, `lastName`.

9. To search by the last name, `Cruise` in our case, enter `http://localhost:8080/people/search/findByLastName?lastName=Cruise` in the browser's address bar.

10. We will now update the last name and age of the person with ID 1; Tom Cruise it is for now. Let's update the last name to `Hanks` and age to `58`. To do this, we will use the HTTP `PATCH` request, and the request will be sent to `http://localhost:8080/people/1`, which uniquely identifies the document to update. The body of the HTTP `PATCH` request is `{"lastName":"Hanks", "age":58}`. Refer to the following screenshot for the request we sent out for update:

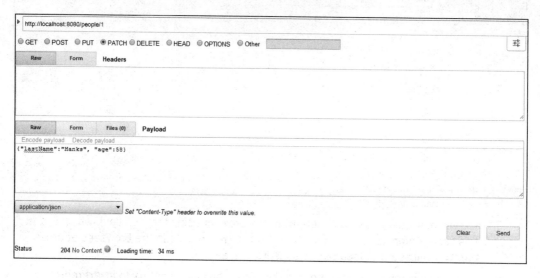

11. To validate whether our update went through successfully or not (we know it did as we got a response status 204 after the `PATCH` request), enter `http://localhost:8080/people/1` again in the browser's address bar.

12. Finally, we will delete the document. This is straightforward, and we will simply send a `DELETE` request to `http://localhost:8080/people/1`. Once the `DELETE` request is successful, send an HTTP GET request from the browser to `http://localhost:8080/people/1`, and we will not get any document in return.

How it works...

We will not be reiterating the `spring-data-mongodb` concepts in this recipe, but we will look at some of the annotations we added specifically for the REST interface to the repository class. The first one is on top of the class name as follows:

```
@RepositoryRestResource(path="people")
public interface PersonRepository extends PagingAndSortingRepository<Person, Integer> {
```

This is used to instruct the server that this CRUD repository can be accessed using the `people` resource. This is the reason why we always make HTTP `GET` and `POST` requests on `http://localhost:8080/people/`.

The second annotation is in the `findByLastName` method. We have the following method signature:

```
Person findByLastName(@Param("lastName") String lastName);
```

Here, the `lastName` method parameter is annotated with the `@Param` annotation, which is used to annotate the name of the parameter that will have the value of the `lastName` parameter that will be passed while invoking this method on the repository. If we look at step 9 in the previous section, we will see that `findByLastName` is invoked using an HTTP `GET` request, and the value of the URL parameter, `lastName`, is used as the string value passed while invoking the repository method.

Our example here is pretty simple with just one parameter used for the search operation. We can have multiple parameters for the repository method and, accordingly, an equal number of parameters in the HTTP request, which will be mapped to these parameters for the method to be invoked on the CRUD repository. For some data type such as dates to be sent out, use the `@DateTimeFormat` annotation, which will be used to specify the date and time format. For more information on this annotation and its usage, refer to the Spring Javadoc at `http://docs.spring.io/spring/docs/current/javadoc-api/`.

That was all about the `GET` request we make to the REST interface to query and search data. Initially, we created document data sending an HTTP `POST` request to the server. To create new documents, we will always send a `POST` request with the document to be created as a body of the request to the URL that identifies the REST endpoint, in our case, `http://localhost:8080/people/`. All documents posted to this collection will make use of `PersonRepository` to persist a person in the corresponding collection.

Our final three steps were to update and delete the person. The HTTP request types to perform these operations are `PATCH` and `DELETE`, respectively. In step 10, we updated the document for the person, Tom Cruise, and updated his last name and age. To achieve this, our `PATCH` request was sent to a URL `http://localhost:8080/people/1`; this URL identifies a specific person instance. Note that, when we wanted to create a new person, our `POST` request was always sent to `http://localhost:8080/people` as against the `PATCH` and `DELETE` requests, where we sent the HTTP request to a URL that represents the specific person we want to update or delete. In the case of update, the body of the `PATCH` request is a JSON document whose provided fields will replace the corresponding fields in the target document to update. All the other fields will be left as they are. In our case, `lastName` and `age` of the target document were updated, and `firstName` was left untouched. In the case of delete, the message body was not empty, and the `DELETE` request itself instructs that the target to which the request was sent should be deleted.

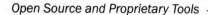

You might also send a PUT request instead of PATCH to a URL that identifies a specific person; in this case, the entire document in the collection will get updated or replaced with the document provided as part of the PUT request.

See also

- The spring-data-rest home page at http://projects.spring.io/spring-data-rest/, where you can find links to its Git repository, reference manual, and Javadoc URLs

Installing the GUI-based client, MongoVUE, for MongoDB

In this recipe, we will look at a GUI-based client for MongoDB. Throughout the book, we have used the Mongo shell to perform various operations we need. Its advantages are as follows:

- It comes packaged with the MongoDB installation
- Being lightweight, you need not worry about it taking up your system's resources
- On servers where GUI-based interfaces are not present, the shell is the only option to connect, query, and administer the server instance

Having said this, if you are not on a server and want to connect to a database instance to query, view the plan of a query, administer, and so on, it is nice to have a GUI with these features to let you do things at a click of a button. As a developer, we always query our relational database with a GUI-based thick client; so why not do the same for MongoDB?

In this recipe, we will see how to install some features of a MongoDB client, MongoVUE. This client is only available on Windows machines. This product has both a paid version (with various levels of licensing per number of users) and a free version that has some limitations. For this recipe, we'll look at the free version.

Getting ready

For this recipe, the following steps are necessary:

1. Start a single instance of the MongoDB server. The port on which the connections are accepted will be the default one, 27017.

2. Import the following two collections from the command prompt after the MongoDB server is started:

   ```
   $ mongoimport --type json personTwo.json -c personTwo -d test
   --drop
   ```

   ```
   $ mongoimport --type csv -c postalCodes -d test pincodes.csv
   --headerline --drop
   ```

How to do it...

1. Download the installer ZIP for the MongoVUE from `http://www.mongovue.com/downloads/`. Once downloaded, it is a matter of a few clicks and the software gets installed.
2. Open the installed application. As this is a free version, we will have all the features available for the first 14 days, after which some of the features will not be available. The details of this can be seen at `http://www.mongovue.com/purchase/`.
3. The first thing we will do is add a database connection as follows:
   1. Once the following window is opened, click on the **+** button to add a new connection.

Open Source and Proprietary Tools

2. Once opened, we will get another window in which we will fill in the server-connection details. Fill in the following details in the new window and click on **Test**. This should succeed if the connection works. Finally, click on **Save**, as shown in the following screenshot:

3. Once added, connect to the instance.

4. In the left-hand-side navigation panel, we will see the instances added and the databases in them, as shown in the following screenshot:

As we see in the preceding screenshot, hovering the mouse over the name of the collection shows us the size and count of the documents in the collection

5. Let's see how to query a collection and get all the documents. We will use the `postalCodes` collection for our test. Right-click on the collection name and then click on **View**. We will see the contents of the collection shown as **Tree View**, where we can expand and see the contents; **Table View**, which shows the contents in a tabular grid; or **Text View**, which shows the contents as normal JSON text.

6. Let's see what happens when we query a collection with nested documents; `personTwo` is a collection with the following sample document in it:

```
{
  "_id" : 1,
  "_class" : "com.packtpub.mongo.cookbook.domain.Person2",
  "firstName" : "Steve",
  "lastName" : "Johnson",
  "age" : 30,
  "gender" : "Male",
  "residentialAddress" : {
    "addressLineOne" : "20, Central street",
    "city" : "Sydney",
    "state" : "NSW",
    "country" : "Australia"
  }
}
```

7. When we query to see all the documents in the collection, we will see the following screenshot:

_id	_class	firstName	lastName	age	gender	residentialAddress
1	com.packtpub...	Steve	Johnson	30	Male	{4 Keys}
2	com.packtpub...	Amit	Sharma	25	Male	{5 Keys}
3	com.packtpub...	Neha	Sharma	27	Female	{5 Keys}

3 Documents (0 to 2)

8. The **residentialAddress** column shows that the value is a nested document with the given number of fields present in it. Hovering your mouse over it shows the nested document. Alternatively, you can click on the column to show the contents in this document again as a grid. Once the nested document(s) are shown, you can click on the top of the grid to come back one level.

Open Source and Proprietary Tools

Let's see how to write queries to retrieve the selected documents:

1. Right-click on the **postalCodes** collection and click on **Find**. We will type the following query in the **{Find }** textbox and in the **{Sort}** field, and click on the **Find** button:

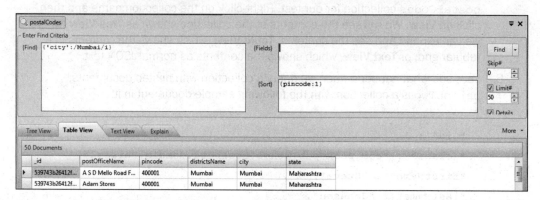

2. We can choose from the tab the type of view we want, such as **Tree View**, **Table View**, or **Text View**. The plan of the query is also shown. Whenever any operation is run, the **Learn** shell at the bottom shows the actual Mongo query executed. In this case, we will see the following query:

   ```
   [ 11:17:07 PM ]
   db.postalCodes.find({ "city" : /Mumbai/i }).limit(50);
   db.postalCodes.find({ "city" : /Mumbai/i }).limit(50).explain();
   ```

3. The plan of a query is also shown every time and, as of the current version 1.6.9.0, there is no way to disable the setting that shows the query plan with the query.

4. In the **Tree View** tab, right-clicking on a document will give you more options such as expanding it, copying the JSON contents, adding keys to this document, and removing the document. Try to remove a document from this collection with a right-click and also try adding any additional keys to the document.

5. You might choose to restore the documents by reimporting the data from the `postalCodes` collection.

To insert a document in the collection, perform the following steps. We will insert a document in the `personTwo` collection.

1. Right-click on the **personTwo** collection name and click on **Insert/Import Documents...**, as shown in the following screenshot:

Chapter 9

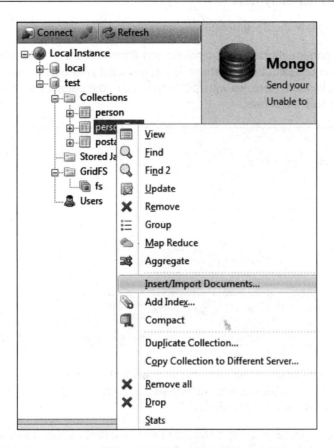

2. Another pop-up window will come up where you can choose to enter a single JSON document or a valid text file with JSON documents to be imported. We will import the following document by importing a single document:

   ```
   {
     "_id" : 4,
     "firstName" : "Jack",
     "lastName" : "Jones",
     "age" : 35,
     "gender" : "Male"
   }
   ```

3. Query the collection once the document is imported successfully; we will see the newly imported document along with the old ones.

Open Source and Proprietary Tools

Let's see how to update the document:

1. You can either right-click on the collection name on the left-hand side and click on **Update** or select the **Update** option at the top. In either case, we will have the following window. Here, we will update the age of the person we inserted in the previous step as follows:

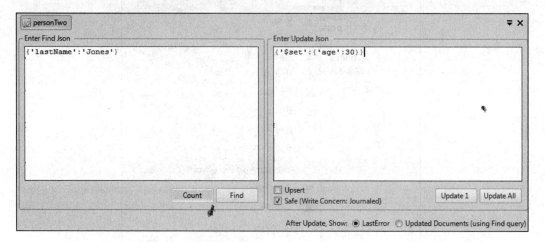

Some things to note in this GUI are the query textbox on the left-hand side to find the document to be updated, and the update JSON on the right-hand side, which will be applied to the selected document(s).

2. Before you update, you might choose to hit the **Count** button to see the number of documents that can be updated (in this case, one). On clicking on **Find**, we can see the documents in the tree form. On the right-hand side, below the update JSON text, we have the option to update one document and multiple documents by clicking on **Update 1** or **Update All**, respectively.

    - You might choose whether the `Upsert` operation is to be done or not if the document(s) for the given find condition are not found
    - The radio buttons in the bottom-right corner of the screen, as shown in the preceding screenshot, either show the output of the `getLastError` operation or the result after update; in this case, a query will be executed to find the document(s) updated
    - However, the `find` query is not foolproof and might return different results from those truly updated as a separate query, the same as in the **{Find}** textbox, and the `update` and `find` operations are not atomic

We have queried on small collections so far. As the size of the collection increases, queries that perform full collection scans are not acceptable, and we need to create indexes as follows:

1. To create an index by `lastName` in the ascending order and `age` in the descending order for instance, we will invoke `db.personTwo.ensureIndex({'lastName':1, 'age':-1})`.
2. Using MongoVUE, there is a way to visually create an index by right-clicking on the collection name on the left-hand side of the screen and selecting **Add Index...**.
3. In the new pop-up window, enter the name of the index and select the **Visual** tab, as shown in the following screenshot. Select the **lastName** and **age** fields with the **Ascending (1)** and **Descending (-1)** values, respectively.

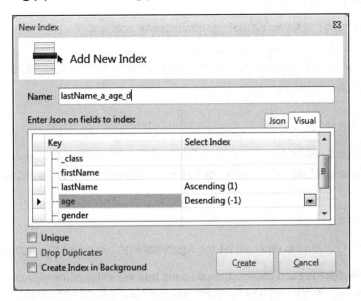

4. Once the preceding details are filled in, click on **Create**. This will create the index for us by firing the **ensureIndex** command, as we can see in the Learn Shell below.
5. You can opt for the index to be unique and drop duplicates (which will be enabled when **Unique** is selected) or even create big, long-running index creations in the background. Note the **Json** tab next to the **Visual** tab. That is the place where you can type in the **ensureIndex** command as you do from the shell to create the index.

Now, we will see how to drop an index:

1. Simply expand the tree on the left-hand side.
2. On expanding the collection, we will see all the indexes created on it.
3. Except for the default index on the `_id` field, all other indexes can be dropped.
4. Simply right-click on the name and select **Drop index** to drop or click on **Properties** to view its properties.

Open Source and Proprietary Tools

After seeing how to do the basic CRUD operations and after creating the index, let's look at how to execute the aggregation operations.

1. There are no visual tools such as index creation in case of aggregation but simply a text area where we can enter our aggregation pipeline. In the following sample, we will perform aggregation on the `postalCodes` collection to find the top five states by the number of times they appear in the collection:

   `{'$project' : {'state':1, '_id':0}},`

   `{'$group': {'_id':'$state', 'count':{'$sum':1}}},`

   `{'$sort':{'count':-1}},`

   `{'$limit':5}`

2. We will have the following aggregation pipeline entered:

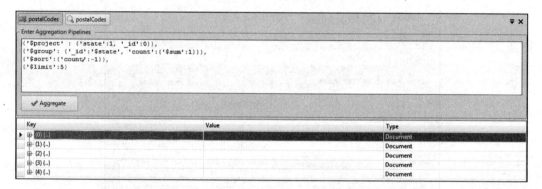

3. Once the pipeline is entered, hit the **Aggregate** button to get the aggregation results.

Executing MapReduce is even cooler. The use case that we will execute is similar to the one we used earlier, but we will see how to implement the MapReduce operation using MongoVUE:

1. To execute a MapReduce job, right-click on the collection name in the left-hand-side menu and click on **Map Reduce**.

2. This option is right above the **Aggregate** option, as seen in the previous screenshot. This gives us a pretty neat GUI to enter the **Map**, **Reduce**, **Finalize**, and **In & Out** options, as shown in the following screenshot:

Chapter 9

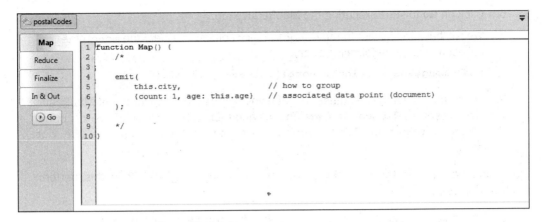

3. The `Map` function is simply as follows:

   ```
   function Map() {
     emit(this.state, 1)
   }
   ```

4. The `Reduce` function is as follows:

   ```
   function Reduce(key, values) {
     return Array.sum(values)
   }
   ```

5. Leave the **Finalize** method unimplemented and in the **In & Out** section, fill in the following details:

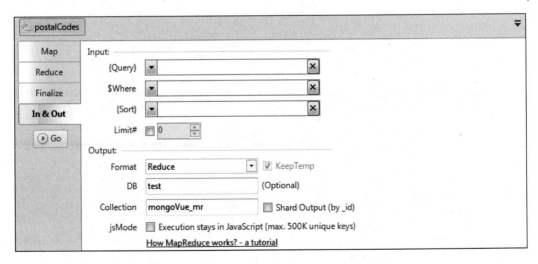

Open Source and Proprietary Tools

6. Click on **Go** to start executing the MapReduce job.

7. We will have the output to the `mongoVue_mr` collection. Query the `mongoVue_mr` collection using the following query:

   ```
   > db.mongoVue_mr.find().sort({value:-1}).limit(5)
   ```

8. Verify the results against those we got using aggregation. The format of MapReduce was chosen as **Reduce**. For more options and their behavior, visit `http://docs.mongodb.org/manual/reference/command/mapReduce/#mapreduce-out-cmd`.

Monitoring the server instances is now easily possible using MongoVUE. To do this, perform the following steps:

1. To monitor an instance, navigate to **Tools | Monitoring** in the top menu.

2. By default, no server will be added, and we will have to click on **+ Add Server** to add a server instance.

3. Select the local instance added or any server you want to monitor and click on **Connect**.

4. We will see quite a lot of monitoring details. MongoVUE uses the `db.serverStatus` command to serve these stats. Thus, to limit the frequency at which we execute this command on busy server instances, we can choose **Refresh Interval** at the top of the screen, as shown in the following screenshot:

Chapter 9

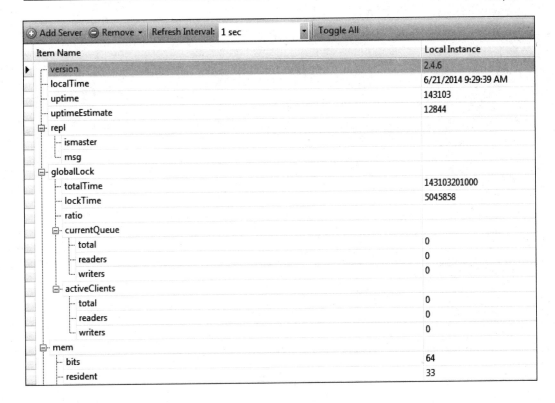

How it works...

What we covered in the previous sections was pretty straightforward; using this information, we can perform the majority of our activities as a developer and administrator.

See also

- *Chapter 4, Administration*
- *Chapter 6, Monitoring and Backups*
- `http://www.mongovue.com/tutorials/` for various tutorials on MongoVUE

Concepts for Reference

This appendix contains some additional information that will help you understand the recipes better. We will discuss write concern and read preference in as much detail as possible.

Write concern and its significance

Write concern is the minimum guarantee that the MongoDB server provides with respect to the write operation done by the client. There are various levels of write concern that are set by the client application, to get a guarantee from the server that a certain stage will be reached in the write process on the server side.

The stronger the requirement for a guarantee, the greater the time taken (potentially) to get a response from the server. With write concern, we don't always need to get an acknowledgement from the server about the write operation being completely successful. For some less crucial data such as logs, we might be more interested in sending more writes per second over a connection. On the other hand, when we are looking to update sensitive information, such as customer details, we want to be sure of the write being successful (consistent and durable); data integrity is crucial and takes precedence over the speed of the writes.

Concepts for Reference

An extremely useful feature of write concern is the ability to compromise between one of the factors: the speed of write operations and the consistency of the data written, on a case-to-case basis. However, it needs a deep understanding of the implications of setting up a particular write concern. The following diagram runs from the left and goes to the right, and shows the increasing level of write guarantees:

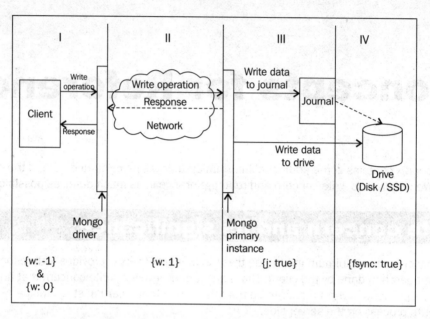

As we move from **I** to **IV**, the guarantee for the performed write gets stronger and stronger, but the time taken to execute the write operation is longer from a client's perspective. All write concerns are expressed here as JSON objects, using three different keys, namely, `w`, `j`, and `fsync`. Additionally, another key called `wtimeout` is used to provide timeout values for the write operation. Let's see the three keys in detail:

- `w`: This is used to indicate whether to wait for the server's acknowledgement or not, whether to report write errors due to data issues or not, and about the data being replicated to secondary. Its value is usually a number and a special case where the value can be `majority`, which we will see later.
- `j`: This is related to journaling and its value can be a Boolean (true/false or 1/0).
- `fsync`: This is a Boolean value and is related to whether the write should wait till the data is flushed to disk or not before responding.
- `wtimeout`: This specifies the timeout for write operations, whereby the driver throws an exception to the client if the server doesn't respond back in seconds within the provided time. We will see the option in some detail soon.

In part **I**, which we have demarcated till driver, we have two write concerns, namely, {w:-1} and {w:0}. Both these write concerns are common, in a sense that they neither wait for the server's acknowledgement upon receiving the write operation, nor do they report any exception on the server side caused by unique index violation. The client will get an ok response and will discover the write failure only when they query the database at some later point of time and find the data missing. The difference is in the way both these respond on the network error. When we set {w:-1}, the operation doesn't fail and a write response is received by the user. However, it will contain a response stating that a network error prevented the write operation from succeeding and no retries for write must be attempted. On the other hand, with {w:0}, if a network error occurs, the driver might choose to retry the operation and throw an exception to the client if the write fails due to network error. Both these write concerns give a quick response back to the invoking client at the cost of data consistency. These write concerns are ok for use cases such as logging, where occasional log write misses are fine. In older versions of MongoDB, {w:0} was the default write concern if none was mentioned by the invoking client. At the time of writing this book, this has changed to {w:1} by default and the option {w:0} is deprecated.

In part **II** of the diagram, which falls between the driver and the server, the write concern we are talking about is {w:1}. The driver waits for an acknowledgement from the server for the write operation to complete. Note that the server responding doesn't mean that the write operation was made durable. It means that the change just got updated into the memory, all the constraints were checked, and any exception will be reported to the client, unlike the previous two write concerns we saw. This is a relatively safe write concern mode, which will be fast, but there is still a slim chance of the data being lost if it crashes in those few milliseconds when the data was written to the journal from the memory. For most use cases, this is a good option to set. Hence, this is the default write concern mode.

Moving on, we come to part **III** of the diagram, which is from the entry point into the server as far as the journal. The write concern we are looking for here is at {j:1} or {j:true}. This write concern ensures a response to the invoking client only when the write operation is written to the journal. What is a journal though? This is something that we saw in depth in *Chapter 4, Administration*, but for now, we will just look at a mechanism that ensures that the writes are made durable and the data on the disk doesn't get corrupted in the event of server crashes.

Finally, let's come to part **IV** of the diagram; the write concern we are talking about is {fsync:true}. This requires that the data be flushed to disk to get before sending the response back to the client. In my opinion, when journaling is enabled, this operation doesn't really add any value, as journaling ensures data persistence even on server crash. Only when journaling is disabled does this option ensure that the write operation is successful when the client receives a success response. If the data is really important, journaling should never be disabled in the first place as it also ensures that the data on the disk doesn't get corrupted.

Concepts for Reference

We have seen some basic write concerns for a single-node server or those relevant to the primary node only in a replica set.

 An interesting thing to discuss is, what if we have a write concern such as {w:0, j:true}? We do not wait for the server's acknowledgement and also ensure that the write has been made to the journal. In this case, journaling flag takes precedence and the client waits for the acknowledgement of the write operation. One should avoid setting such ambiguous write concerns to avoid unpleasant surprises.

We will now talk about write concern when it involves secondary nodes of a replica set as well. Let's take a look at the following diagram:

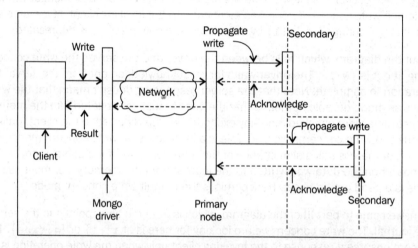

Any write concern with a w value greater than one indicates that secondary nodes too need to acknowledge before sending a response back. As seen in the preceding diagram, when a primary node gets a write operation, it propagates that operation to all secondary nodes. As soon as it gets a response from a predetermined number of secondary nodes, it acknowledges the client that the write has been successful. For example, when we have a write concern {w:3}, it means that the client should be sent a response only when three nodes in the cluster acknowledge the write. These three nodes include the primary node. Hence, it is now down to two secondary nodes to respond back for a successful write operation.

Appendix

However, there is a problem with providing a number for the write concern. We need to know the number of nodes in the cluster and accordingly set the value of w. A low value will send an acknowledgement to a few nodes replicating the data. A value too high may unnecessarily slow the response back to the client, or in some cases, might not send a response at all. Suppose you have a three-node replica set and we have {w:4} as the write concern, the server will not send an acknowledgement till the data is replicated to three secondary nodes, which do not exist as we have just two secondary nodes. Thus, the client waits for a very long time to hear from the server about the write operation. There are a couple of ways to address this problem:

- Use the wtimeout key and specify the timeout for the write concern. This will ensure that a write operation will not block for longer than the time specified (in milliseconds) for the wtimeout field of the write concern. For example, {w:3, wtimeout:10000} ensures that the write operation will not block more than 10 seconds (10,000 ms), after which an exception will be thrown to the client. In the case of Java, a WriteConcernException will be thrown with the root cause message stating the reason as timeout. Note that this exception does not rollback the write operation. It just informs the client that the operation did not get completed in the specified amount of time. It might later be completed on the server side, some time after the client receives the timeout exception. It is up to the application program to deal with the exception and programmatically take the corrective steps. The message for the timeout exception does convey some interesting details, which we will see on executing the test program for the write concern.

- A better way to specify the value of w, in the case of replica sets, is by specifying the value as majority. This write concern automatically identifies the number of nodes in a replica set and sends an acknowledgement back to the client when the data is replicated to a majority of nodes. For example, if the write concern is {w:"majority"} and the number of nodes in a replica set is three, then majority will be 2. Whereas, at the later point in time, when we change the number of nodes to five, the majority will be 3 nodes. The number of nodes to form a majority automatically gets computed when the write concern's value is given as majority.

Now, let us put the concepts we discussed into use and execute a test program that will demonstrate some of the concepts we just saw.

Setting up a replica set

To set up a replica set, you should know how to start the basic replica set with three nodes. Refer to the *Starting multiple instances as part of a replica set* recipe in *Chapter 1, Installing and Starting the MongoDB Server*. This recipe is built on that recipe because it needs an additional configuration while starting the replica set, which we will discuss in the next section. Note that the replica used here has a slight change in configuration to the one you have used earlier.

Concepts for Reference

Here, we will use a Java program to demonstrate various write concerns and their behavior. The *Connecting to a single node from a Java client* recipe in *Chapter 1, Installing and Starting the MongoDB Server*, should be visited until Maven is set up. This can be a bit inconvenient if you are coming from a non-Java background.

>
> The Java project named `Mongo Java` is available for download at the book's website. If the setup is complete, you can test the project just by executing the following command:
> ```
> mvn compile exec:java
> -Dexec.mainClass=com.packtpub.mongo.cookbook.
> FirstMongoClient
> ```
> The code for this project is available for download at the book's website. Download the project named `WriteConcernTest` and keep it on a local drive ready for execution.

So, let's get started:

1. Prepare the following configuration file for the replica set. This is identical to the config file that we saw in the *Starting multiple instances as part of a replica set* recipe in *Chapter 1, Installing and Starting the MongoDB Server*, where we set up the replica set, as follows, with just one difference, `slaveDelay:5, priority:0`:

    ```
    cfg = {
        _id:'repSetTest',
        members:[
            {_id:0, host:'localhost:27000'},
            {_id:1, host:'localhost:27001'},
            {_id:2, host:'localhost:27002', slaveDelay:5, priority:0}
        ]
    }
    ```

2. Use this config to start a three-node replica set, with one node listening to port `27000`. The others can be any ports of your choice, but stick to `27001` and `27002` if possible (we need to update the config accordingly if we decide to use a different port number). Also, remember to set the name of the replica set as `replSetTest` for the `replSet` command-line option while starting the replica set. Give some time to the replica set to come up before going ahead with next step.

3. At this point, the replica set with the earlier mentioned specifications should be up and running. We will now execute the test code provided in Java, to observe some interesting facts and behaviors of different write concerns. Note that this program also tries to connect to a port where no Mongo process is listening for connections. The port chosen is `20000`; ensure that before running the code, no server is up and running and listening to port `20000`.

Appendix

4. Go to the root directory of the `WriteConcernTest` project and execute the following command:

   ```
   mvn compile exec:java
   -Dexec.mainClass=com.packtpub.mongo.cookbook.WriteConcernTests
   ```

   It should take some time to execute completely, depending on your hardware configuration. Roughly around 35 to 40 seconds were taken on my machine, which has a spinning disk drive with a 7200 RPM.

Before we continue analyzing the logs, let us see what those two additional fields added to the config file to set up the replica were. The `slaveDelay` field indicates that the particular slave (the one listening on port `27002` in this case) will lag behind the primary by 5 seconds. That is, the data being replicated currently on this replica node will be the one that was added on to the primary 5 seconds ago. Secondly, this node can never be a primary and hence, the `priority` field has to be added with the value `0`. We have already seen this in detail in *Chapter 4, Administration*.

Let us now analyze the output from the preceding command's execution. The Java class provided need not be looked at here; the output on the console is sufficient. Some of the relevant portions of the output console are as follows:

```
[INFO] --- exec-maven-plugin:1.2.1:java (default-cli) @ mongo-cookbook-wctest ---
Trying to connect to server running on port 20000
Trying to write data in the collection with write concern {w:-1}
Error returned in the WriteResult is NETWORK ERROR
Trying to write data in the collection with write concern {w:0}
Caught MongoException.Network trying to write to collection, message is Write operation to server localhost/127.0.0.1:20000 failed on database test
Connected to replica set with one node listening on port 27000 locally

Inserting duplicate keys with {w:0}
No exception caught while inserting data with duplicate _id
Now inserting the same data with {w:1}
Caught Duplicate Exception, exception message is { "serverUsed" : "localhost/127.0.0.1:27000" , "err" : "E11000 duplicate key error index: test.writeConcernTest.$_id_  dup key: { : \"a\" }" , "code" : 11000 , "n" : 0 , "lastOp" : { "$ts" :1386009990 , "$inc" : 2} , "connectionId" : 157 , "ok" : 1.0}
Average running time with WriteConcern {w:1, fsync:false, j:false} is 0 ms
Average running time with WriteConcern {w:2, fsync:false, j:false} is 12 ms
Average running time with WriteConcern {w:1, fsync:false, j:true} is 40 ms
```

Concepts for Reference

```
Average running time with WriteConcern {w:1, fsync:true, j:false} is 44 ms
Average running time with WriteConcern {w:3, fsync:false, j:false} is 5128
ms
Caught WriteConcern exception for {w:5}, with following message {
"serverUsed" : "localhost/127.0.0.1:27000" , "n" : 0 , "lastOp" : {
"$ts" : 1386009991 , "$inc" : 18} , "connectionId" : 157 , "wtimeout"
: true , "waited" : 1004 , "writtenTo" : [ { "_id" : 0 , "host" :
"localhost:27000"} , { "_id" : 1 , "host" : "localhost:27001"}] , "err" :
"timeout" , "ok" : 1.0}
    [INFO] ------------------------------------------------------------
------
[INFO] BUILD SUCCESS
[INFO] ------------------------------------------------------------
-----
[INFO] Total time: 36.671s
[INFO] Finished at: Tue Dec 03 00:16:57 IST 2013
[INFO] Final Memory: 13M/33M
[INFO] ------------------------------------------------------------
-----
```

The first statement in the log states that we try to connect to a Mongo process listening on port 20000. As there should not be a Mongo server running and listening to this port for client connections, all our write operations to this server should not succeed, and this will now give us a chance to see what happens when we use the write concerns {w:-1} and {w:0} and write to this nonexistent server.

The next two lines in the output show that when we have the write concern {w:-1}, we do get a write result back, but it contains the error flag set to indicate a network error. However, no exception is thrown. In the case of the write concern {w:0}, we do get an exception in the client application for any network errors. Of course, all other write concerns ensuring a strict guarantee will throw an exception in this case too.

Now we come to the portion of the code that connects to the replica set where one of the nodes is listening to port 27000 (if not, the code will show the error on the console and terminate). Now, we attempt to insert a document with a duplicate _id field ({'_id':'a'}) into a collection, once with the write concern {w:0} and once with {w:1}. As we see in the console, the former ({w:0}) didn't throw an exception and the insert went through successfully from the client's perspective, whereas the latter ({w:1}) threw an exception to the client, indicating a duplicate key. The exception contains a lot of information about the server's hostname and port, at the time when the exception occurred: the field for which the unique constraint failed; the client connection ID; error code; and the value that was not unique and caused the exception. The fact is that, even when the insert was performed using {w:0} as the write concern, it failed. However, as the driver didn't wait for the server's acknowledgement, it was never communicated about the failure.

Moving on, we now try to compute the time taken for the write operation to complete. The time shown here is the average of the time taken to execute the same operation with a given write concern five times. Note that these times will vary on different instances of execution of the program, and this method is just meant to give some rough estimates for our study. We can conclude from the output that the time taken for the write concern {w:1} is less than that of {w:2} (asking for an acknowledgement from one secondary node) and the time taken for {w:2} is less than {j:true}, which in turn is less than {fsync:true}. The next line of the output shows us that the average time taken for the write operation to complete is roughly 5 seconds when the write concern is {w:3}. Any guesses on why that is the case? Why does it take so long? The reason is, when w is 3, we send an acknowledgement to the client only when two secondary nodes acknowledge the write operation. In our case, one of the nodes is delayed from the primary by about 5 seconds, and thus, it can acknowledge the write only after 5 seconds, and hence, the client receives a response from the server in roughly 5 seconds.

Let us do a quick exercise here. What do you'll think would be the approximate response time when we have the write concern as {w:'majority'}? The hint here is, for a replica set of three nodes, two is the majority.

Finally we see a timeout exception. Timeout is set using the wtimeout field of the document and is specified in milliseconds. In our case, we gave a timeout of 1000 ms, that is 1 second, and the number of nodes in the replica set to get an acknowledgement from before sending the response back to the client is 5 (four secondary instances). Thus, we have the write concern as {w:5, wtimeout:1000}. As our maximum number of nodes is three, the operation with the value of w set to 5 will wait for a very long time until two more secondary instances are added to the cluster. With the timeout set, the client returns and throws an error to the client, conveying some interesting details. The following is the JSON sent as an exception message:

```
{ "serverUsed" : "localhost/127.0.0.1:27000" , "n" : 0 , "lastOp" : {
"$ts" : 1386015030 , "$inc" : 1} , "connectionId" : 507 , "wtimeout"
: true , "waited" : 1000 , "writtenTo" : [ { "_id" : 0 , "host" :
"localhost:27000"} , { "_id" : 1 , "host" : "localhost:27001"}] , "err" :
"timeout" , "ok" : 1.0}
```

Let us look at the interesting fields. We start with the n field. This indicates the number of documents updated. As in this case it is an insert and not an update, it stays 0. The wtimeout and waited fields tell us whether the transaction did timeout and the amount of time for which the client waited for a response; in this case 1000 ms. The most interesting field is writtenTo. In this case, the insert was successful on these two nodes of the replica set when the operation timed out, and hence, it is seen in the array. The third node has a slaveDelay value of 5 seconds and, hence, the data is still not written to it. This proves that the timeout doesn't roll back the insert and it does go through successfully. In fact, the node with slaveDelay will also have the data after 5 seconds, even if the operation times out, and this makes perfect sense as it keeps the primary and secondary instances in sync. It is the responsibility of the application to detect such timeouts and handle them.

Concepts for Reference

Read preference for querying

In the previous section, we saw what a write concern is and how it affects the write operations (insert, update, and delete). In this section, we will see what a read preference is and how it affects query operations. We'll discuss how to use a read preference in separate recipes, to use specific programming language drivers.

When connected to an individual node, query operations will be allowed by default when connected to a primary, and in case if it is connected to a secondary node, we need to explicitly state that it is ok to query from secondary instances by executing `rs.slaveOk()` from the shell.

However, consider connecting to a Mongo replica set from an application. It will connect to the replica set and not a single instance from the application. Depending on the nature of the application, it might always want to connect to a primary; always to a secondary; prefer connecting to a primary node but would be ok to connect to a secondary node in some scenarios and vice versa and finally, it might connect to the instance geographically close to it (well, most of the time).

Thus, the read preference plays an important role when connected to a replica set and not to a single instance. In the following table, we will see the various read preferences that are available and what their behavior is in terms of querying a replica set. There are five of them and the names are self-explanatory:

Read preference	Description
`primary`	This is the default mode and it allows queries to be executed only on primary instances. It is the only mode that guarantees the most recent data, as all writes have to go through a primary instance. Read operations however will fail if no primary is available, which happens for a few moments when a primary goes down and continues till a new primary is chosen.
`primaryPreferred`	This is identical to the preceding primary read preference, except that during a failover, when no primary is available, it will read data from the secondary and those are the times when it possibly doesn't read the most recent data.
`secondary`	This is exactly the opposite to the default primary read preference. This mode ensures that read operations never go to a primary and a secondary is chosen always. The chances of reading inconsistent data that is not updated to the latest write operation are maximal in this mode. It, however, is ok (in fact, preferred) for applications that do not face end users and are used for some instances to get hourly statistics and analytics jobs used for in-house monitoring, where the accuracy of the data is least important, but not adding a load to the primary instance is key. If no secondary instance is available or reachable, and only a primary instance is, the read operation will fail.

Appendix

Read preference	Description
`secondaryPreferred`	This is similar to the preceding secondary read preference, in all aspects except that if no secondary is available, the read operations will go to the primary instance.
`nearest`	This, unlike all the preceding read preferences, can connect either to a primary or a secondary. The primary objective for this read preference is minimum latency between the client and an instance of a replica set. In the majority of the cases, owing to the network latency and with a similar network between the client and all instances, the instance chosen will be one that is geographically close.

Similar to how write concerns can be coupled with shard tags, read preferences can also be used along with shard tags. As the concept of tags has already been introduced in *Chapter 4, Administration*, you can refer to it for more details.

We just saw what the different types of read preferences are (except for those using tags) but the question is, how do we use them? We have covered Python and Java clients in this book and will see how to use them in their respective recipes. We can set read preferences at various levels: at the client level, collection level, and query level, with the one specified at the query level overriding any other read preference set previously.

Let us see what the nearest read preference means. Conceptually, it can be visualized as something like the following diagram:

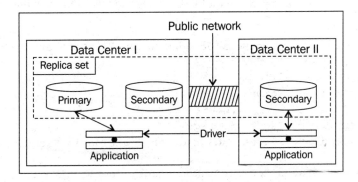

Concepts for Reference

A Mongo replica set is set up with one secondary, which can never be a primary, in a separate data center and two (one primary and a secondary) in another data center. An identical application deployed in both the data centers, with a primary read preference, will always connect to the primary instance in **Data Center I**. This means, for the application in **Data Center II**, the traffic goes over the public network, which will have high latency. However, if the application is ok with slightly stale data, it can set the read preference as the nearest, which will automatically let the application in **Data Center I** connect to an instance in **Data Center I** and will allow an application in **Data Center II** to connect to the secondary instance in **Data Center II**.

But then the next question is, how does the driver know which one is the nearest? The term "geographically close" is misleading; it is actually the one with the minimum network latency. The instance we query might be geographically further than another instance in the replica set, but it can be chosen just because it has an acceptable response time. Generally, better response time means geographically closer.

The following section is for those interested in internal details from the driver on how the nearest node is chosen. If you are happy with just the concepts and not the internal details, you can safely skip the rest of the contents.

Knowing the internals

Let us see some pieces of code from a Java client (driver 2.11.3 is used for this purpose) and make some sense out of it. If we look at the `com.mongodb.TaggableReadPreference.NearestReadPreference.getNode` method, we see the following implementation:

```
@Override
ReplicaSetStatus.ReplicaSetNode getNode(ReplicaSetStatus.ReplicaSet set) {
  if (_tags.isEmpty())
    return set.getAMember();

  for (DBObject curTagSet : _tags) {
    List<ReplicaSetStatus.Tag> tagList = getTagListFromDBObject(curTagSet);
    ReplicaSetStatus.ReplicaSetNode node = set.getAMember(tagList);
    if (node != null) {
      return node;
    }
  }
  return null;
}
```

For now, if we ignore the contents where tags are specified, all it does is execute `set.getAMember()`.

Appendix

The name of this method tells us that there is a set of replica set members and we returned one of them randomly. Then what decides whether the set contains a member or not? If we dig a bit further into this method, we see the following lines of code in the `com.mongodb.ReplicaSetStatus.ReplicaSet` class:

```
public ReplicaSetNode getAMember() {
   checkStatus();
   if (acceptableMembers.isEmpty()) {
      return null;
   }
   return acceptableMembers.get(random.nextInt(acceptableMembers.size()));
}
```

Ok, so all it does is pick one from a list of replica set nodes maintained internally. Now, the random pick can be a secondary, even if a primary can be chosen (because it is present in the list). Thus, we can now say that when the nearest is chosen as a read preference, and even if a primary is in the list of contenders, it might not necessarily be chosen randomly.

The question now is, how is the `acceptableMembers` list initialized? We see it is done in the constructor of the `com.mongodb.ReplicaSetStatus.ReplicaSet` class as follows:

```
this.acceptableMembers =
   Collections.unmodifiableList(calculateGoodMembers(all,
      calculateBestPingTime(all, true),
      acceptableLatencyMS, true));
```

The `calculateBestPingTime` line just finds the best ping time of all (we will see what this ping time is later).

Another parameter worth mentioning is `acceptableLatencyMS`. This gets initialized in `com.mongodb.ReplicaSetStatus.Updater` (this is actually a background thread that updates the status of the replica set continuously), and the value for `acceptableLatencyMS` is initialized as follows:

```
slaveAcceptableLatencyMS = Integer.parseInt(System.getProperty("com.mongodb.slaveAcceptableLatencyMS", "15"));
```

As we can see, this code searches for the system variable called `com.mongodb.slaveAcceptableLatencyMS`, and if none is found, it initializes to the value 15, which is 15 ms.

Concepts for Reference

This `com.mongodb.ReplicaSetStatus.Updater` class also has a `run` method that periodically updates the replica set stats. Without getting too much into it, we can see that it calls `updateAll`, which eventually reaches the `update` method in `com.mongodb.ConnectionStatus.UpdatableNode`:

```
long start = System.nanoTime();
CommandResult res = _port.runCommand(_mongo.getDB("admin"),
  isMasterCmd);
long end = System.nanoTime()
```

All it does is execute the `{isMaster:1}` command and record the response time in nanoseconds. This response time is converted to milliseconds and stored as the ping time. So, coming back to the `com.mongodb.ReplicaSetStatus.ReplicaSet` class it stores, all `calculateGoodMembers` does is find and add the members of a replica set that are no more than `acceptableLatencyMS` milliseconds more than the best ping time found in the replica set.

For example, in a replica set with three nodes, the ping times from the client to the three nodes (node 1, node 2, and node 3) are 2 ms, 5 ms, and 150 ms respectively. As we see, the best time is 2 ms and hence, node 1 goes into the set of good members. Now, from the remaining nodes, all those with a latency that is no more than `acceptableLatencyMS` more than the best, which is *2 + 15 ms = 17 ms*, as 15 ms is the default that will be considered. Thus, node 2 is also a contender, leaving out node 3. We now have two nodes in the list of good members (good in terms of latency).

Now, putting together all that we saw on how it would work for the scenario we saw in the preceding diagram, the least response time will be from one of the instances in the same data center (from the programming language driver's perspective in these two data centers), as the instance(s) in other data centers might not respond within 15 ms (the default acceptable value) more than the best response time due to public network latency. Thus, the acceptable nodes in **Data Center I** will be two of the replica set nodes in that data center, and one of them will be chosen at random, and for **Data Center II**, only one instance is present and is the only option. Hence, it will be chosen by the application running in that data center.

Index

Symbols

$text operator
 URL 223
-d option 229
--fields option 229
-m option 228
-n option 228
-t option 228

A

advanced packaging tool (apt) 72
aggregation operations, in MongoDB
 executing, with Java client 98, 99
 executing, with PyMongo 86, 87
Amazon EC2
 MongoDB, setting up with
  MongoDB AMI 278-289
 MongoDB, setting up without
  MongoDB AMI 289-291
Amazon Elastic MapReduce (Amazon EMR)
 MapReduce job, running on 308-314
 URL 315
Amazon Machine Image. *See* **AMI**
Amazon Web Service (AWS) 267
AMI
 about 278
 URL 278
ar|aw column 120
atomic counters
 implementing, in MongoDB 183, 184

B

background index creation 53-56

backups
 managing, in MMS backup service 259-264
binary data
 storing, in MongoDB 192, 193
buildIndexes option 154
bulk inserts
 URL 326

C

capped collection
 about 188
 normal collection, converting to 191, 192
capped collection cursors
 creating, in MongoDB 188-190
 tailing, in MongoDB 188-190
chunks
 migrating 170-172
 splitting 170-172
collection
 duplicate data, creating 57-59
 renaming 104, 105
 unique indexes, creating 57-59
collection behavior
 modifying, with collMod command 143-145
collection stats
 fields 108, 109
 viewing 106-110
collMod command
 used, for modifying
  collection behavior 143-145
command column 119
command-line options
 --config 10
 --configsvr 11
 --dbpath 10

-f 10
-h 10
--help 10
--logappend 10
--logpath 10
--oplogSize 11
--port 10
--quiet 10
--replSet 11
--shardsvr 11
--smallfiles 11
-v 10
--verbose 10
used, for starting single node instance 9-12
config database
 exploring, in sharded setup 175-177
config file options
 used, for single node installation
  of MongoDB 12, 13
conn column 120
covered indexes
 about 51
 using 52

D

data
 backing up, with out-of-the box tools
  in MongoDB 250-253
 deleting, from Mongo shell 46, 47
 restoring, with out-of-the box tools
  in MongoDB 250-253
 storing, to GridFS from Java client 198-201
 storing, to GridFS from
  Python client 202-204
 updating, from Mongo shell 46, 47
database stats
 viewing 110-113
data file preallocation
 disabling 114
default shard
 configuring, for nonsharded
  collections 168-170
delete column 119
delete operations
 executing, with Java client 94-97
 executing, with PyMongo 78-85

document
 expiring after fixed interval,
  TTL index used 63-65
 padding 114-117
domain-driven sharding
 performing, with tags 172-174

E

Elastic Block Store (EBS) 279
Elastic Cloud Compute (EC2) 279
Elasticsearch
 integrating, with MongoDB for
  full-text search 224-229
 URL 224, 229

F

faults column 120
fields, collection stats
 avgObjSize 108
 count 108
 indexSizes 109
 lastExtentSize 108
 nindexes 108
 ns 108
 numExtents 108
 paddingFactor 109
 size 108
 storageSize 108
 totalIndexSize 109
fields, database stats
 avgObjectSize 113
 collections 113
 dataSize 113
 db 113
 fileSize 113
 indexes 113
 indexSize 113
 nsSizeMB 113
 numExtents 113
 objects 113
 storageSize 113
fields, db.currentOp() operation
 client 133
 lockStats 133
 millis 133

nreturned 133
ns 133
nscanned 133
numYields 133
op 133
query 133
responseLength 133
ts 133
fields, operations
 active 127
 client 127
 connectionId 128
 desc 127
 insert 127
 locks 128
 lockStats 128
 msg 128
 ns 127
 numYields 128
 op 127
 opid 127
 progress 128
 query 127
 secs_running 127
 waitingForLock 128
fields parameter 96
findAllAndRemove method 326
findAndModify method 327
findAndRemove method 326
findByAgeBetween method 323
findByAgeGreaterThanEqual method 323
findByAgeGreaterThan method 323
findByFirstNameAndCountry method 324
findByResidentialAddressCountry method 324
find operation
 about 180, 181
 working 182
findPeopleByLastNameLike method 323
first in first out (FIFO) 65
flat plane (2D) geospatial queries
 executing, in MongoDB with geospatial indexes 209-211
flushes column 119
foreground index creation 53-56
full-text search
 implementing, in MongoDB 218-223

MongoDB, integrating with Elasticsearch 224-229

G

GeoJSON
 URL 212
GeoJSON-compliant data, MongoDB
 about 212-215
 working 216
geospatial indexes
 used, for executing flat plane (2D) geospatial queries in MongoDB 209-211
geospatial operators
 URL 217
getmore column 119
Git
 URL 294
GridFS
 used, for storing large data in MongoDB 194-197
GridFS, from Java client
 data, storing to 198-201
GridFS, from Python client
 data, storing to 202-204
groups
 managing, on MMS console 236-238
GUI-based client
 installing, for MongoDB 338-341

H

Hadoop
 about 293
 streaming, used for running MapReduce jobs 304-307
 URL 295
Hadoop MapReduce job
 writing 301-303

I

idx miss % column 120
in-built profiler
 used, for profiling operations 129-133
index creation
 pitfalls, avoiding 53-56
insert column 119

insert operations
 executing, with Java client 90-93
 executing, with PyMongo 73-78
 interprocess security, in MongoDB 141-143
Infrastructure as a Service (IaaS) 198
I/O operations per second (IOPS) 279

J

Java client
 data, storing to GridFS 198-201
 replica set connection,
  for inserting data 29-32
 replica set connection,
  for querying data 29-32
 single node connection, establishing 15-20
 used, for executing aggregation
  operations in MongoDB 98, 99
 used, for executing delete
  operations 93-97
 used, for executing insert operations 90-92
 used, for executing MapReduce
  operations in MongoDB 100, 101
 used, for executing query operations 90-93
 used, for executing update operations 93-97
Java Database Connectivity (JDBC) 320
Javadoc
 URL 329
Java Persistence API (JPA)
 about 317
 used, for accessing MongoDB 329-332
JIRA
 URL 67, 242

L

least recently used (LRU) 124
local database, replica set
 exploring 156-158
locked column 120

M

mapped column 119
MapReduce
 URL 294, 348

MapReduce job
 executing, with mongo-hadoop
  connector 294-300
 running, on Amazon EMR 308-314
 running, on Hadoop with streaming 304-307
MapReduce operations, in MongoDB
 executing, with Java client 100, 101
 executing, with PyMongo 87-90
 implementing, MongoVUE used 346-348
Maven
 URL 16, 19
MMS
 about 231
 groups, managing 236-238
 MongoDB instances, monitoring 239-247
 monitoring agent, setting up 232-235
 monitoring alerts, setting up 248, 249
 signing up 232-235
 URL 232, 247
 users, managing 236-238
MMS backup service
 backups, managing 259-264
 configuring 253-259
modify operation
 about 180, 181
 working 182
Mongo client, options
 -h 15
 --help 15
 --host 15
 -p 15
 --password 15
 --port 15
 --shell 15
 -u 15
 --username 15
mongo-connector
 URL 228
MongoDB
 accessing, Java Persistence
  API used 329-332
 accessing, over REST 333-338
 aggregation operations, executing 346
 aggregation operations, executing with
  Java client 98, 99

aggregation operations, executing with
    PyMongo 86, 87
atomic counters, implementing 183, 184
binary data, storing in 192-194
capped collection cursors, creating 188-190
capped collection cursors, tailing 188-190
data, backing up with out-of-the
    box tools 250-253
data, restoring with out-of-the
    box tools 250-253
document, inserting in collection 342, 343
document, updating 344
full-text search, implementing 218-223
geospatial indexes, used for executing flat
    plane (2D) geospatial queries 209-211
GUI-based client, installing for 338-342
indexes, creating 345
integrating, with Elasticsearch for full-text
    search 224-229
interprocess security 141-143
large data, storing with GridFS 194-197
MapReduce operations, executing with
    Java client 100, 101
MapReduce operations, executing with
    PyMongo 87-89
MongoVUE, installing for 338-342
operations, performing from
    MongoLab GUI 274-278
queries, writing 342
setting up, as Windows Service 145-147
setting up, on Amazon EC2 with
    MongoDB AMI 278-288
setting up, on Amazon EC2 without
    MongoDB AMI 289-291
single node installation 8
single node installation, with config
    file options 12, 13
triggers, implementing with oplog 204-209
URL 70, 229, 294
users, setting up 133-140

MongoDB AMI
used, for setting up MongoDB on
    Amazon EC2 278-288

MongoDB API
URL 93

MongoDB driver
URL 97

MongoDB instances
monitoring, on MMS 239-247

MongoDB Monitoring Service. *See* **MMS**

mongo-hadoop connector
URL 297, 315
used, for executing MapReduce job 294-300

MongoLab
sandbox MongoDB instance,
    setting up 270-273
URL 268

MongoLab account
managing 268-270
setting up 268-270

MongoLab GUI
operations, performing on
    MongoDB from 274-277

Mongo shell
about 69
data, deleting 46, 47
data, updating 46, 47
pagination, performing 43-45
projections, performing 43-45
querying 43-45
shard connection, creating from 37-40
single node connection, with preloaded
    JavaScript 13-15

Mongo's Python client. *See* **PyMongo**

mongostat utility
about 117, 118
working 118-122

MongoTemplate class, methods
findAllAndRemove 326
findAndModify 327
findAndRemove 326
insert 326
remove 325
save 325
updateFirst 326
updateMulti 326

mongotop utility
about 117
working 118-122

MongoVUE
installing, for MongoDB 338-342
URL 339, 349
used, for implementing
    MapReduce operation 346-348

used, for monitoring server instances 348
monitoring alerts
  setting up, on MMS 248, 249

N

nearest, read preference 361
netIn column 120
netOut column 120
nonsharded collections
  default shard, configuring 168-170
normal collection
  converting, to capped collection 191, 192

O

object relational mapping (ORM) 317
operations
  killing 124-129
  profiling, with in-built profiler 129-133
  viewing 124-129
oplog
  about 159
  analyzing 159
  used, for implementing triggers
    in MongoDB 204-209
  working 160-162
options, mongodump utility
  --authenticationDatabase 252
  -c 252
  --collection 252
  -d 252
  --db 252
  --dbpath 252
  --help 251
  -h 251
  --host 251
  -o 252
  --oplog 252
  --out 252
  -p 251
  --password 251
  --port 251
  -u 251
  --username 251
options, MongoDB import utility
  -c 42
  -d 42

  --drop 42
  --headerline 42
  --type 42
options, mongorestore utility
  --dbpath 252
  --drop 252
  --oplogLimit 253
  --oplogReplay 253

P

pagination
  performing, from Mongo shell 43-45
primaryPreferred, read preference 360
primary, read preference 360
projections
  performing, from Mongo shell 43-45
proof of concept (POC) 197
PuTTY
  URL 279
PyMongo
  about 70
  installing 70-73
  URL 71
  used, for executing aggregation operations
    in MongoDB 86, 87
  used, for executing delete operations 79-86
  used, for executing MapReduce operations
    in MongoDB 87-90
  used, for executing query 73-78
  used, for executing update operations 79-85
  used, for inserting operations 73-78
Python
  URL 70
Python client
  data, storing to GridFS 202-204
Python Package Index (PyPI) tool 72

Q

qr|qw column 120
query column 119
query execution time
  improving 50
querying
  performing, from Mongo shell 43-45
query operations
  executing, with Java client 90-93

executing, with PyMongo 73-77
query parameter 96
query plan
  analysis 49
  execution time, improving 50
  improving, with covered indexes 51, 52
  viewing 48, 49

R

RAID
  URL 292
read preference
  about 360-362
  for querying 360
  nearest 361
  primary 360
  primaryPreferred 360
  secondary 360
  secondaryPreferred 361
remove method 325
remove parameter 96
replica set
  about 21
  configuring 21-25, 147-151
  creating 21
  elections 148
  hidden term 153, 154
  index configuration, building 153, 154
  index creation, URL 162
  local database, exploring 156-158
  member, as arbiter 152
  setting up 355-359
  standalone instance, converting to 26
  stepping down, as primary instance 155
  tagged replica sets, building 163
  votes option 153
  WriteConcern 167
replica set connection
  establishing, for inserting data from
    Java client 29-32
  establishing, for querying data from
    Java client 29-32
  establishing, from Mongo shell for
    inserting data 26-28
  establishing, from Mongo shell for
    querying data 26-28

replica set member
  arbiter 152
  priority 152, 153
res column 119
REST
  MongoDB, accessing over 333-338
returnNew parameter 96
rs.stepDown() method 156

S

sandbox MongoDB instance
  setting up, on MongoLab 270-273
save method 325
secondaryPreferred, read preference 361
secondary, read preference 360
server instances
  monitoring, MongoVUE used 348
server-side scripts
  implementing 185-187
sh.addShardTag method 174
shard connection
  creating, for data operations 37-40
  creating, from Mongo shell 37-40
sharded environment
  setting up 32-36
sharded setup
  config database, exploring 175-177
sh.removeShardTag method 174
sh.splitAt function 172
Simple Storage Service (S3) 277, 308
single node connection
  establishing, from Java client 15-20
  establishing, from Mongo shell with preloaded
    JavaScript 13-15
  prerequisites, from Java client 15
single node installation, MongoDB
  steps 8
single node instance
  starting, command-line options used 9-12
slaveDelay option 151, 154
social security number 39
sort parameter 96
sparse indexes
  about 59
  creating 60-62

spherical indexes, MongoDB
  about 212-215
  working 216, 217
spring-data-mongodb
  project, URL 329
  used, for development 318-329
Spring Javadoc
  URL 337
standard error (stderr) 307
stemming process
  about 219
  URL 219
streaming
  used, for running MapReduce jobs on Hadoop 304-307

T

tagged replica sets
  building 163-166
  building, use cases 163
tags
  used, for domain-driven sharding performance 172-174
test data
  creating 41-43
time column 120
Time To Live index. *See* **TTL index**
triggers
  implementing, in MongoDB with oplog 204-209
TTL index
  about 63, 64
  used, for document expiring after fixed interval 63-65
  used, for document expiring at given time 66, 67

U

unique indexes, on collection
  creating 57, 58
update column 119
updateFirst method 326
updateMulti method 326
update operations
  executing, with Java client 94-97
  executing, with PyMongo 78-85
update parameter 96
upsert parameter 96
users
  managing, on MMS console 236-238
  setting up, in MongoDB 133-140

V

VirtualBox
  URL 294
vsize column 119

W

Windows Service
  MongoDB, setting up as 145-147
working set
  estimating 122-124
write concern
  about 351
  fsync key 352
  j key 352
  replica set, setting up 355-359
  significance 351-355
  w key 352
  wtimeout option 352

Thank you for buying
MongoDB Cookbook

About Packt Publishing

Packt, pronounced 'packed', published its first book "*Mastering phpMyAdmin for Effective MySQL Management*" in April 2004 and subsequently continued to specialize in publishing highly focused books on specific technologies and solutions.

Our books and publications share the experiences of your fellow IT professionals in adapting and customizing today's systems, applications, and frameworks. Our solution based books give you the knowledge and power to customize the software and technologies you're using to get the job done. Packt books are more specific and less general than the IT books you have seen in the past. Our unique business model allows us to bring you more focused information, giving you more of what you need to know, and less of what you don't.

Packt is a modern, yet unique publishing company, which focuses on producing quality, cutting-edge books for communities of developers, administrators, and newbies alike. For more information, please visit our website: www.packtpub.com.

About Packt Open Source

In 2010, Packt launched two new brands, Packt Open Source and Packt Enterprise, in order to continue its focus on specialization. This book is part of the Packt Open Source brand, home to books published on software built around Open Source licenses, and offering information to anybody from advanced developers to budding web designers. The Open Source brand also runs Packt's Open Source Royalty Scheme, by which Packt gives a royalty to each Open Source project about whose software a book is sold.

Writing for Packt

We welcome all inquiries from people who are interested in authoring. Book proposals should be sent to author@packtpub.com. If your book idea is still at an early stage and you would like to discuss it first before writing a formal book proposal, contact us; one of our commissioning editors will get in touch with you.

We're not just looking for published authors; if you have strong technical skills but no writing experience, our experienced editors can help you develop a writing career, or simply get some additional reward for your expertise.